写楽園

紅秋 

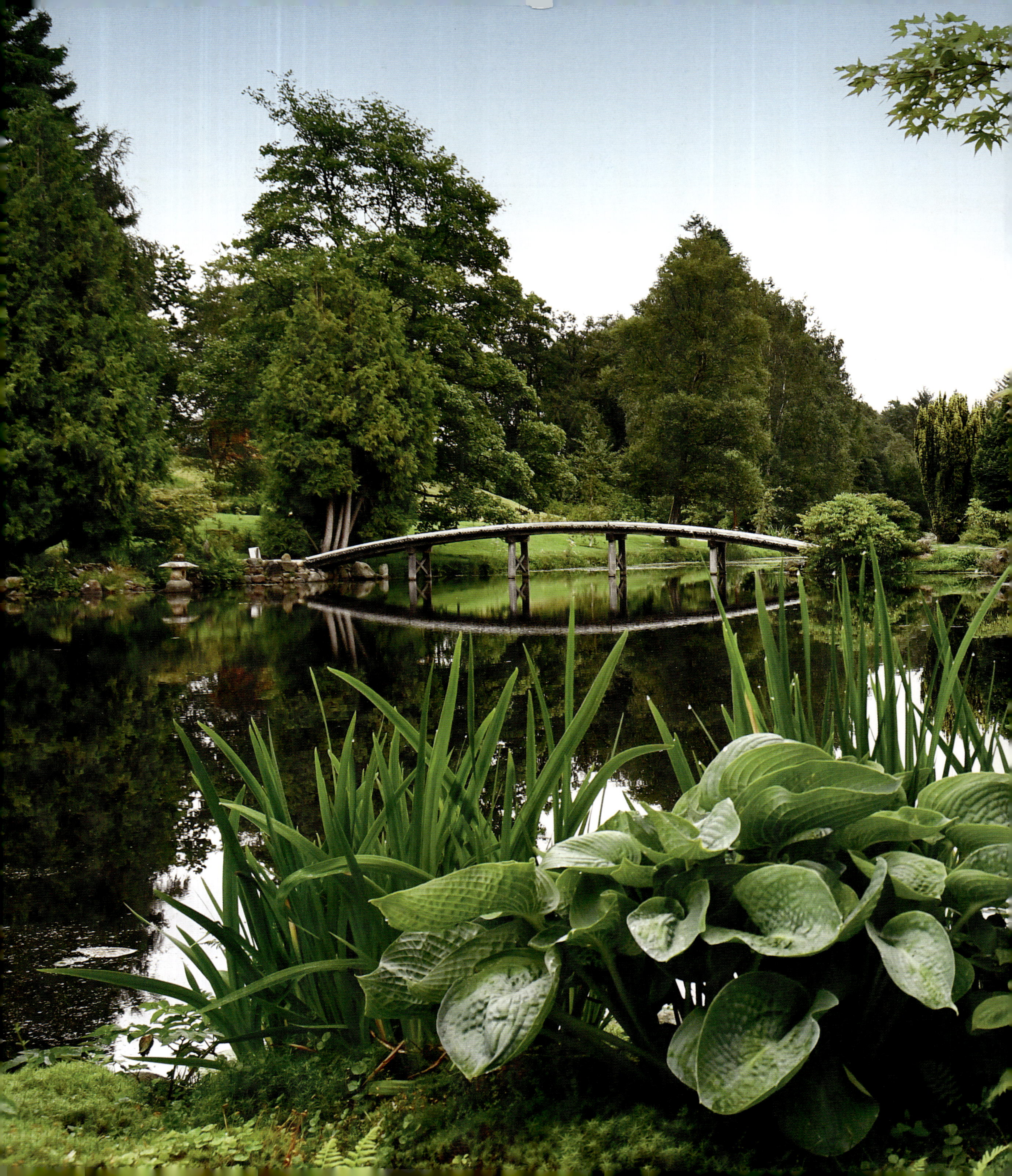

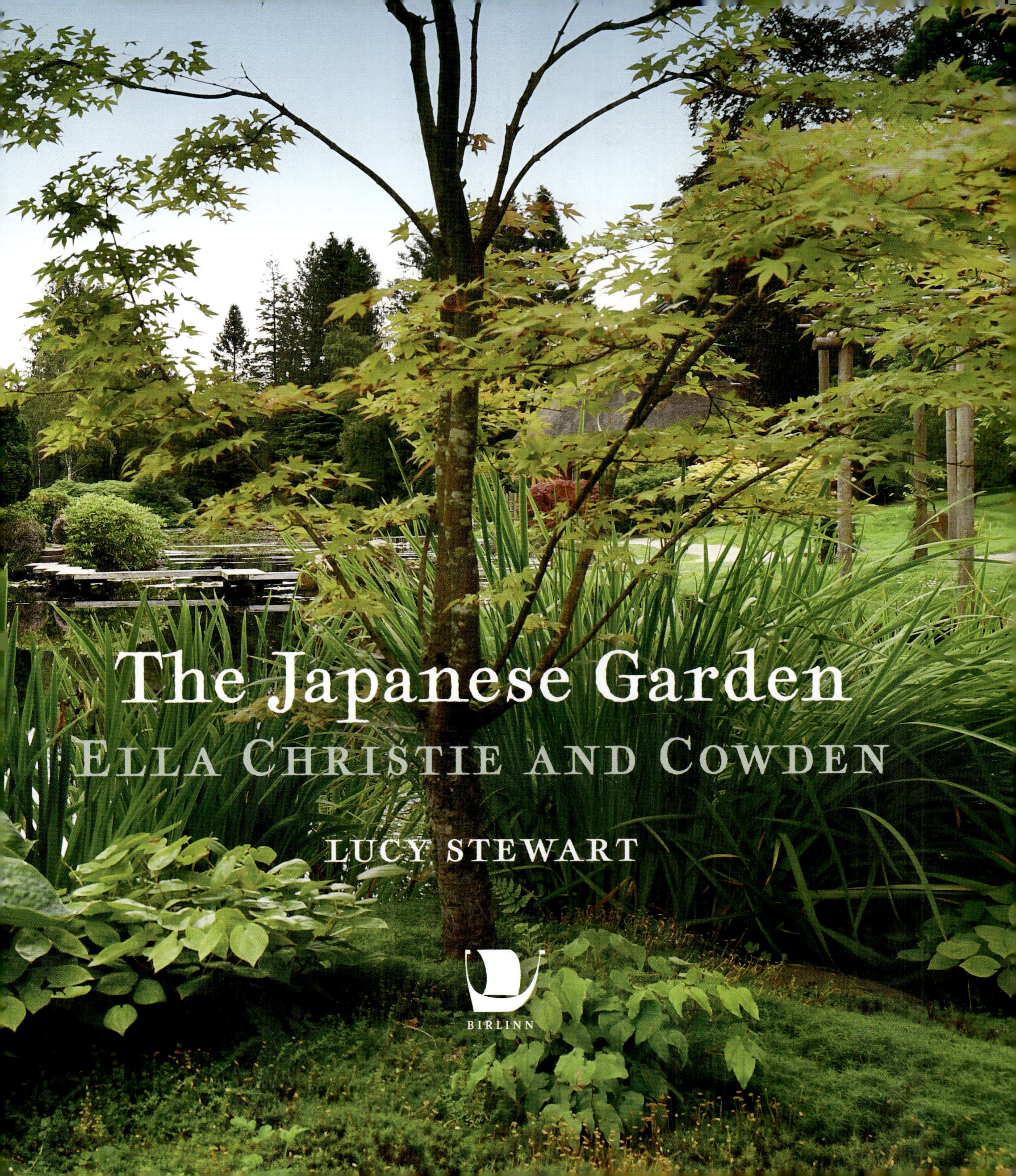

The Japanese Garden

Ella Christie and Cowden

Lucy Stewart

BIRLINN

First published in 2026 by
Birlinn Limited
West Newington House
10 Newington Road
Edinburgh
EH9 1QS

www.birlinn.co.uk
www.cowdengarden.com

Copyright © Lucy Stewart 2026

ISBN 978 1 78027 995 4

British Library Cataloguing-in-Publication Data
A catalogue record of this book is available on request from
the British Library

Designed and typeset by Robert Dalrymple

Papers used by Birlinn are from well-managed forests and
other responsible sources

Printed and bound by Replika Press, India

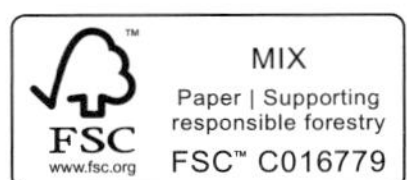

Principal photography by Lucy Stewart; additional images
by Sara Stewart, Molly Stewart, Brian Vass and the Japanese
Garden at Cowden SCIO.

front cover The Garden Pavilion at Cowden, its design based
on the *Shōkin-tei* teahouse at the Katsura Imperial Villa in
Kyoto, Japan.

inside front cover *Tomei ishi* stones to guide the visitor
along the correct path, *tomei ishi* literally translating as
'stop stone'.

back cover View of the garden on 7 September 2024, when a
10th anniversary party was held to celebrate the completion
of the restoration.

inside back cover Scottish frogs at home in a Japanese garden.

on page 1 Calligraphy designed for the Welcome Gate
by Akemi Lucas to announce the name of the garden,
Sharaku-En.

Contents

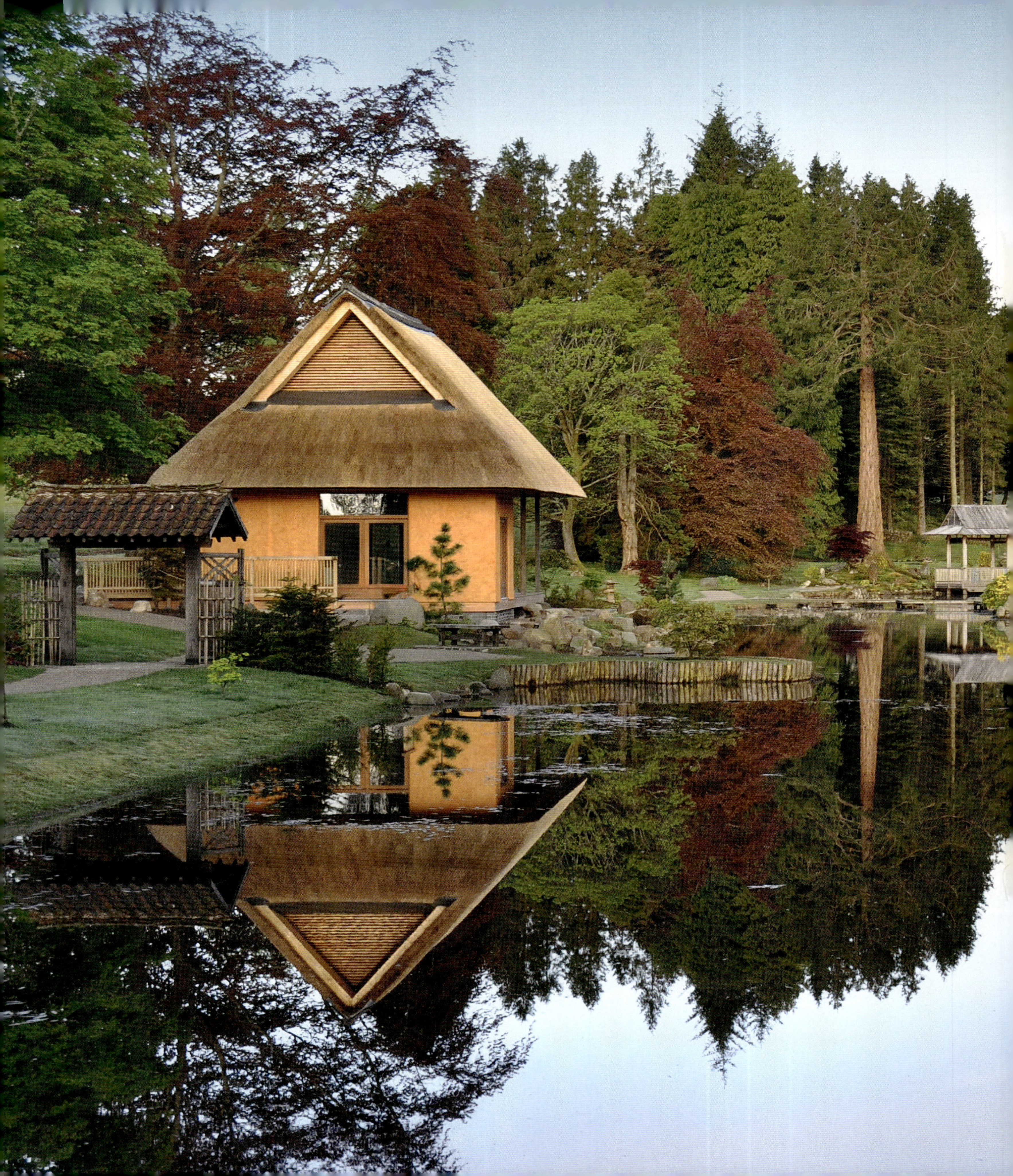

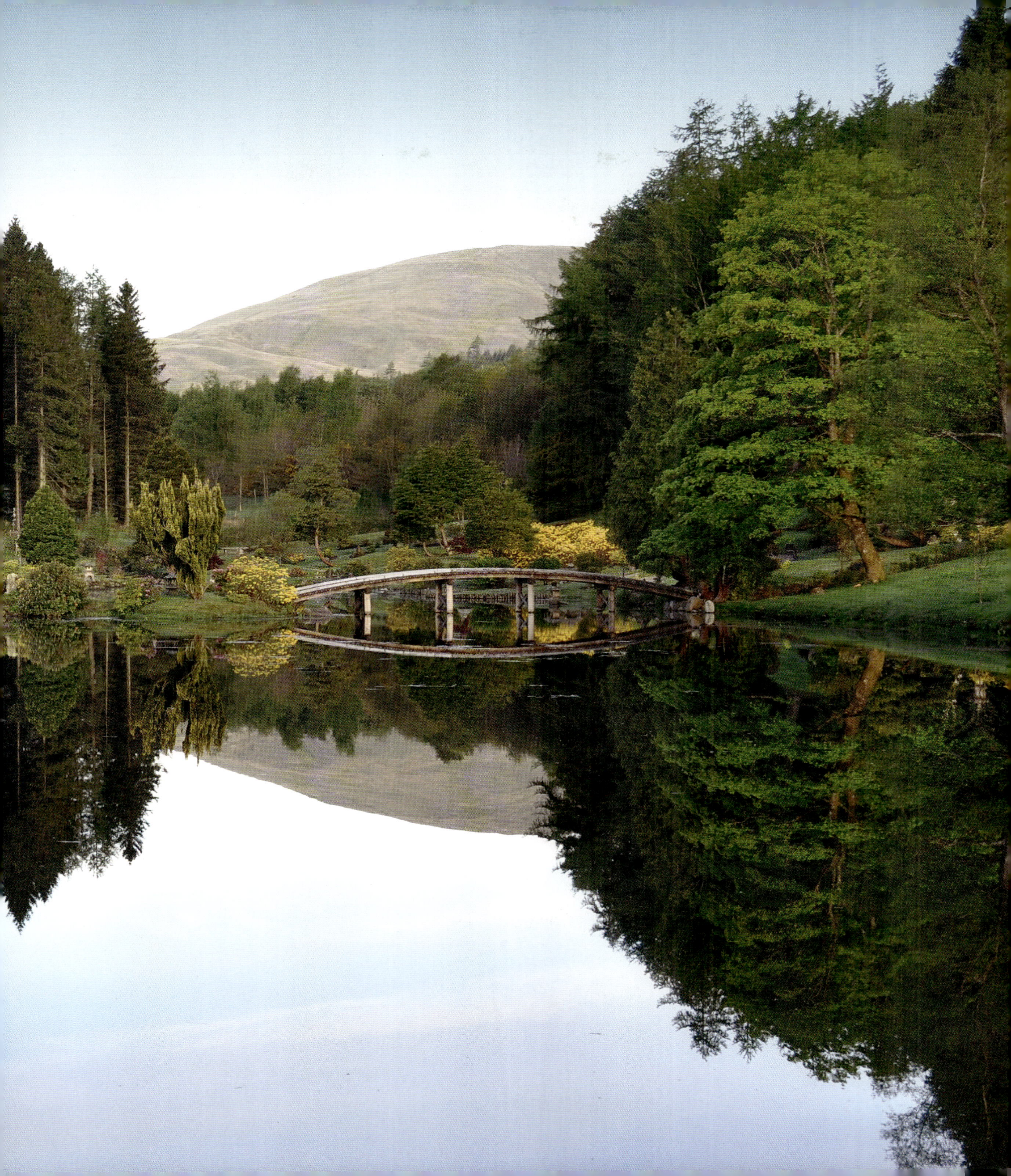

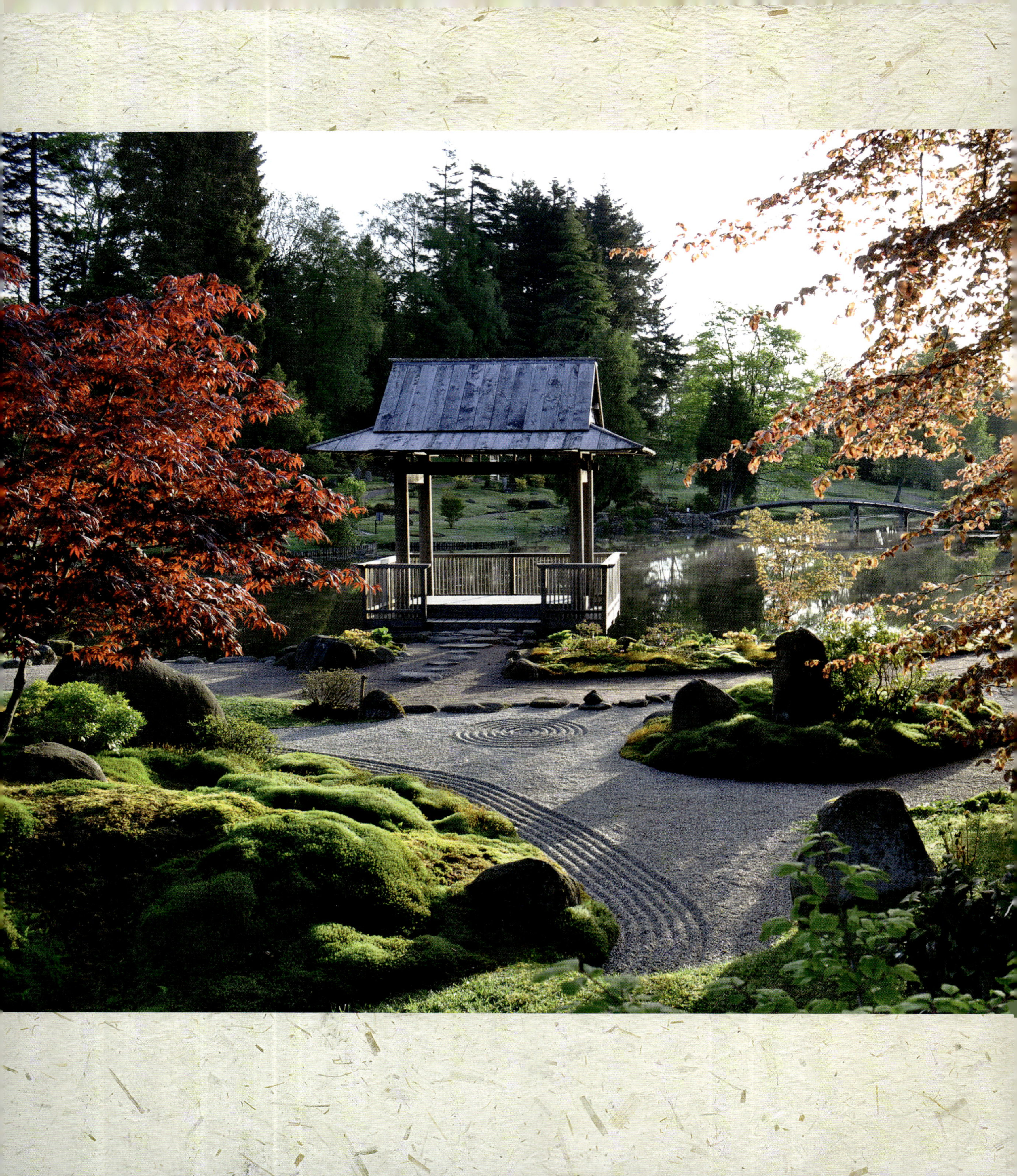

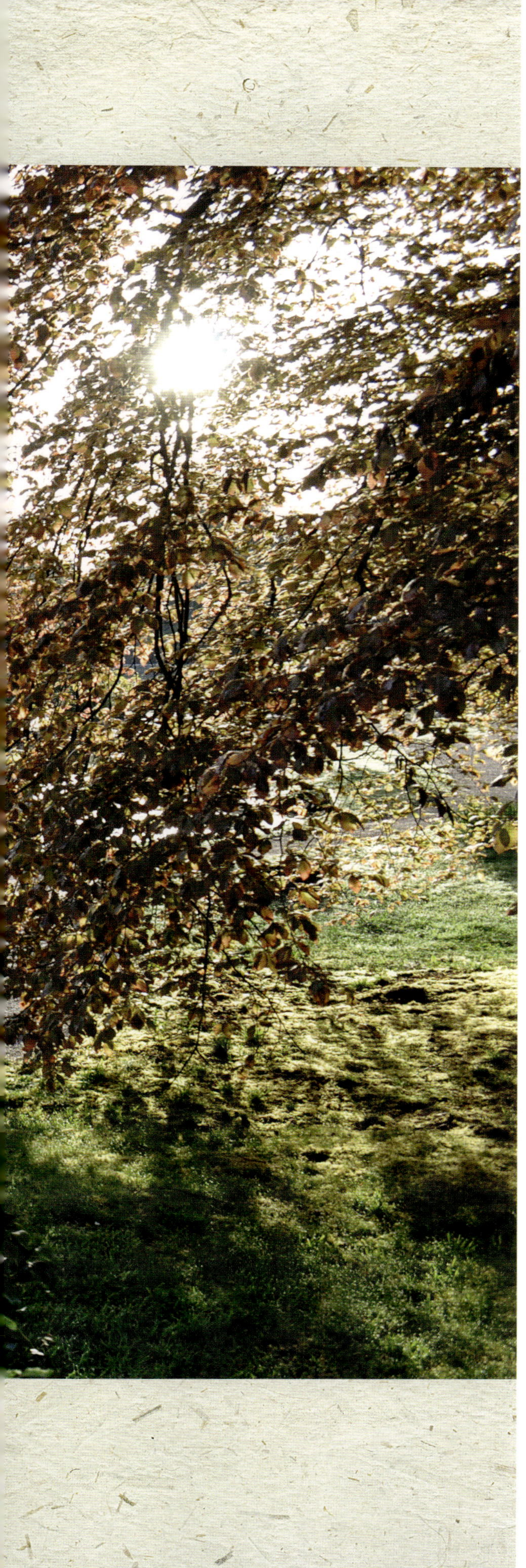

Introduction

ISABELLA CHRISTIE (1861–1949) was one of the great female travellers of her generation. 'A woman Marco Polo,' as the *Sunday Times* described her, she made remarkable journeys across Central Asia, India, China, Japan, Borneo and what were then known as Burma, Malaya and Western Tibet (now Myanmar, Malaysia and Ladakh) and was one of the first female fellows of the Royal Geographical Society. That she was born in the middle of Queen Victoria's reign, when women, even those from privileged backgrounds such as hers, had relatively limited freedoms, makes Ella's story all the more astonishing.

In 1910, having packed 'a Jaeger rug and a cotton padded quilt to act as a mattress and so help relieve the unevenness of a Central Asian hotel bedstead,' Ella travelled across Europe to Istanbul, crossed the Black Sea and the Caspian Sea and followed the Silk Road through Turkestan to Bokhara, Samarkand and Tashkent. Thwarted at the Chinese border, she failed to reach her intended destination of Khiva, but she was never one to give up and returned in 1912. This time, she took a route from St. Petersburg, involving a voyage down the Oxus river and various stages by cart. She eventually reached her goal, becoming the first recorded British woman to have made it to the ancient desert trading post on the Silk Road from Persia. Ella commemorated her arrival with a photograph taken by a guard, in which she is seen standing by a gateway at the city walls. Not since Captain Burnaby in 1875 had a Briton visited Khiva.

Ella's unflagging curiosity made her a great observer of foreign cultures. She was fascinated by how people lived, what they ate, their dress, customs and religious practices and she took many photographs. She also chronicled her journeys in hundreds of letter-journals, a rich source for her biographers. The recipient of these letters was her sister Alice, to whom she was very close. The sisters wrote regularly to each other – Alice also in diary form – and their correspondence constitutes a fascinating social document of a world that, in their own lifetimes, saw the invention of the motor car, the aeroplane, the telephone and electricity, as well as two World Wars. 'Singly, each [sister] was a delightful companion, together, they were a party,' wrote Alice's daughter-in-law, Averil Stewart, in her book *Alicella*. 'With Ella it seemed always to be holiday time even when one was busy. To go abroad with her was to watch the world through her all-seeing eyes.'

Ella's intelligence and sense of fun, as well as her perspicacity, charmed those she met and she had an enviable ability to open doors. Nothing daunted her. One night, she'd be sleeping in a station waiting room, keeping the insects at bay with thick sprinklings of Aragatz powder, the next, staying with the commissioner of a remote province. In Mukden, China, her bed was a wooden board and she subsisted on cheese and walnuts; in India, she dined with Lord Kitchener, the Viceroy and a Maharajah and stayed at Government House in Bombay.

A great shopper wherever she went, Ella was a true Scotswoman when it came to bargaining in the bazaars she so loved to visit. She found it incomprehensible that anyone would accept the initial price offered for a Bokhara carpet, a Turkmen robe, a piece of Chinese porcelain or the services of a guide without a spirited round of haggling. Forthright and determined, she summarily sacked several of her guides, finding that she was more capable than they were at navigating the local terrain.

Part of what makes Ella's life so fascinating is the contrast between her two worlds. At home in Scotland, she was chatelaine of Cowden Castle, Clackmannanshire, which required her not only to manage a large estate, but also to host house parties and patronise many local events. All her foreign purchases found their way back to Cowden, so that visitors might find themselves hanging their hat on a Burmese idol or making a telephone call beneath a frieze of Dyak shields.

A deep sense of duty and social responsibility ran through Ella's veins and, through her life, she was involved in helping many people in a myriad ways. Outstanding in this respect was her work as *directrice* of two canteens on the Western Front – at Bar-sur-Aube and Mulhouse – during and just after the First World War.

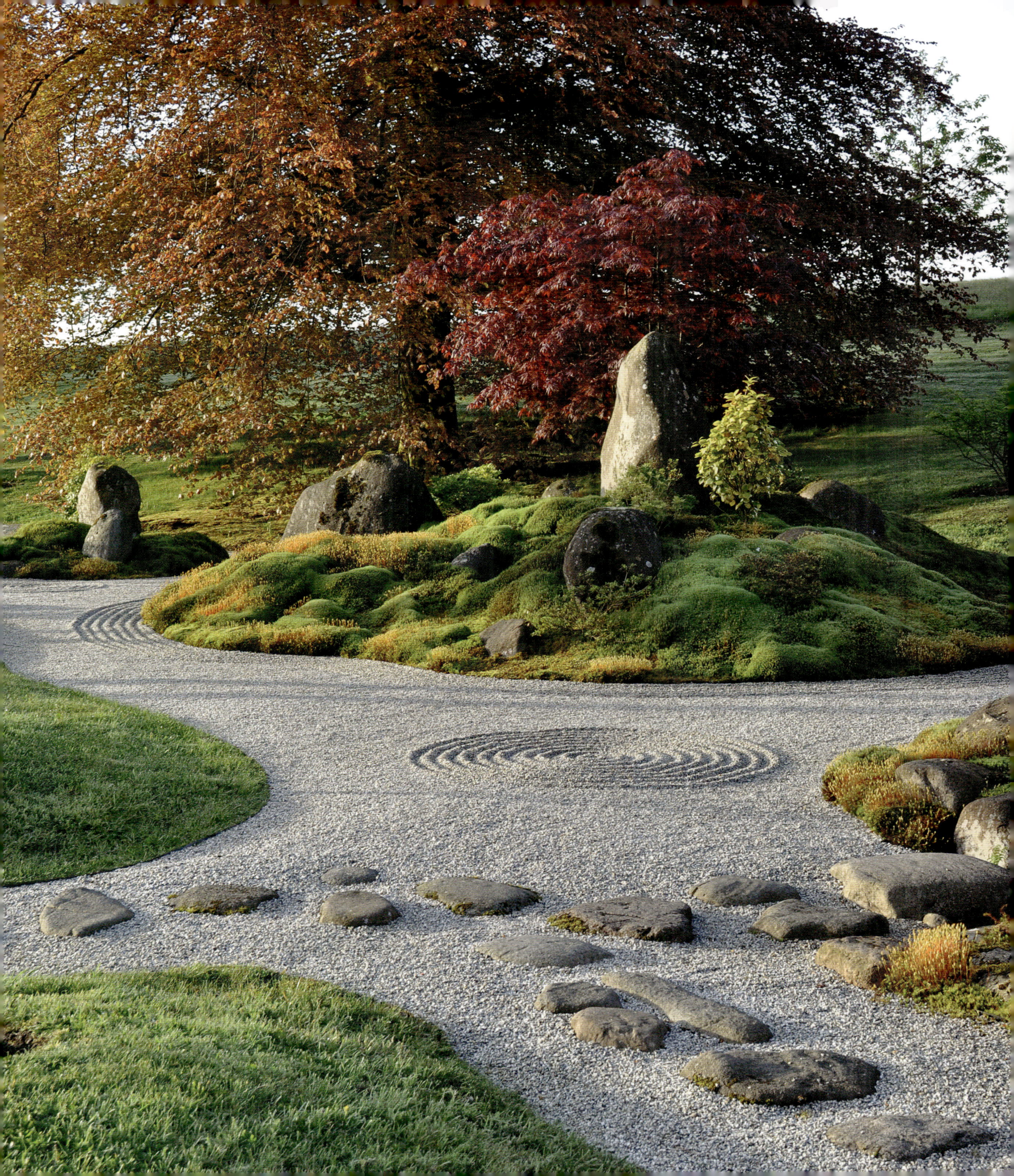

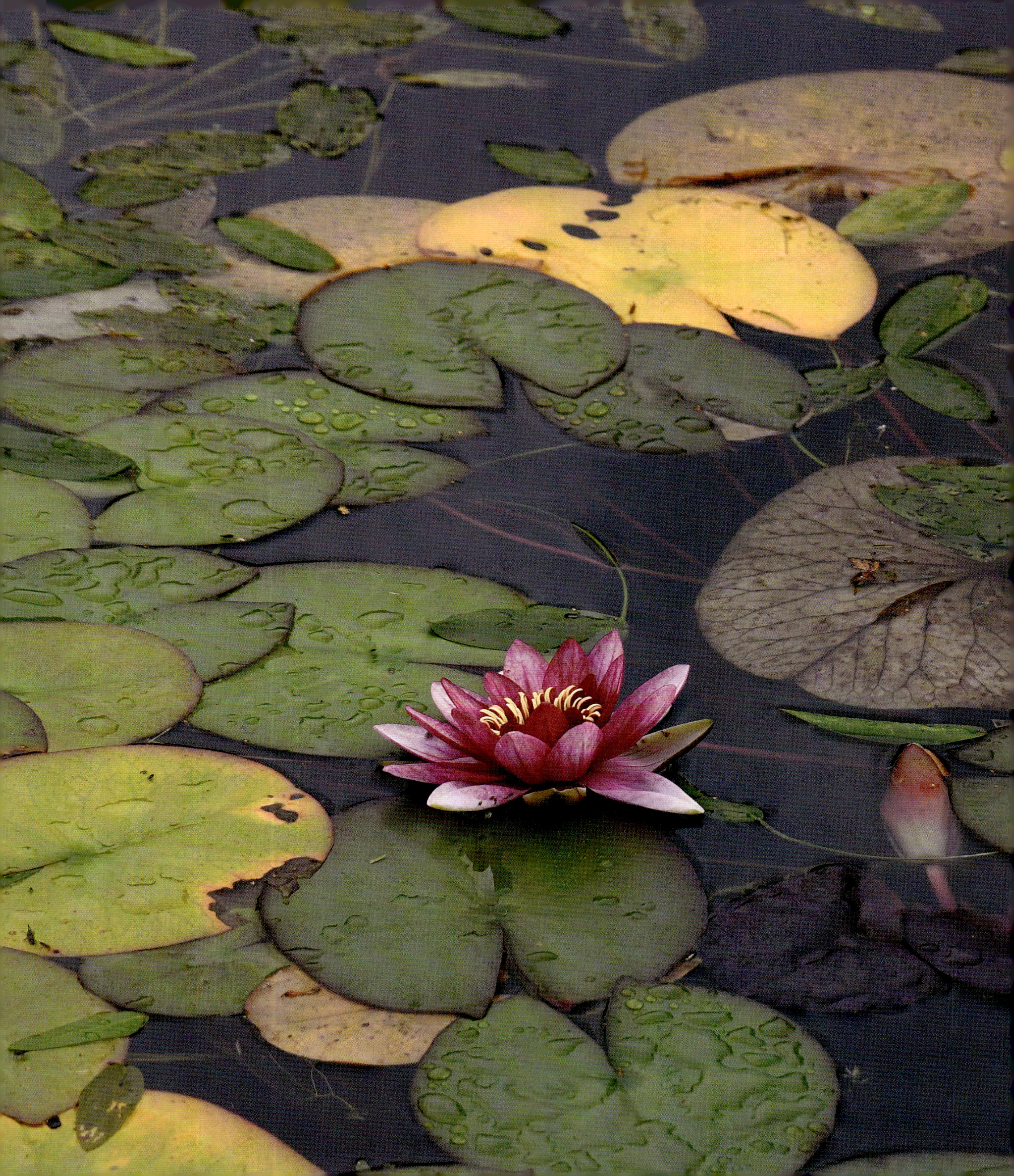

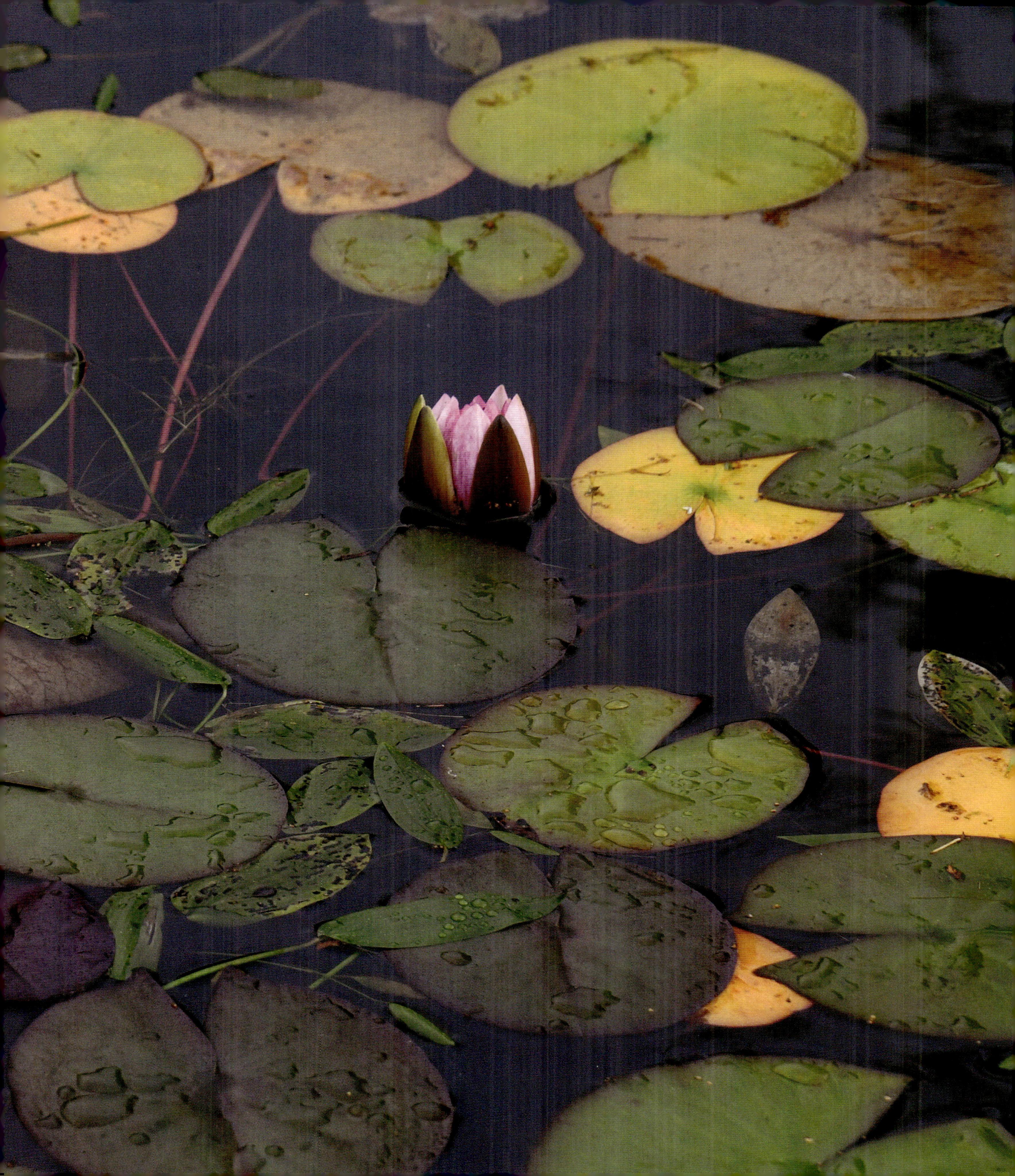

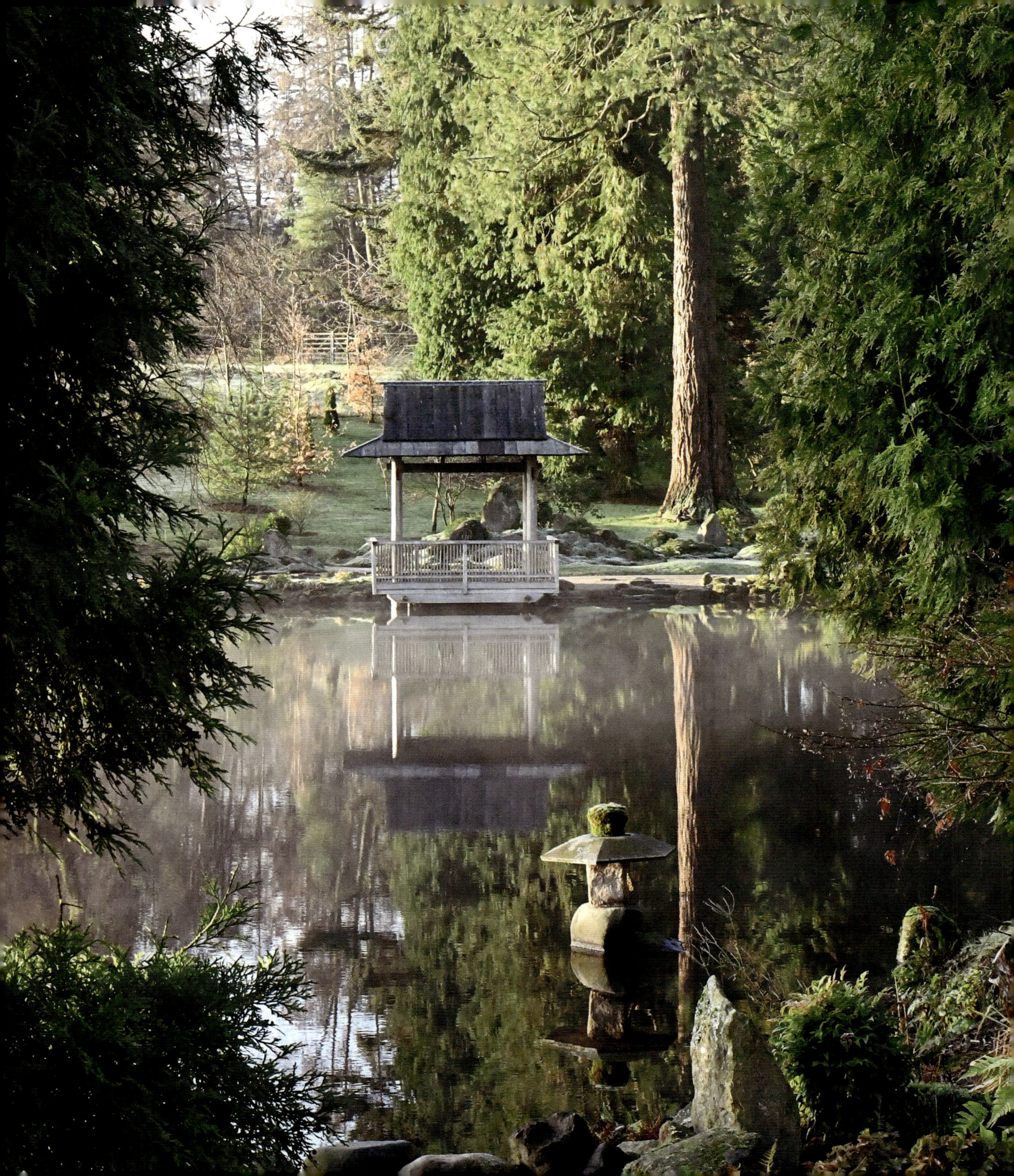

It was a trip to Japan in 1907 that inspired her to build her famous Japanese garden at Cowden. As ever, Ella's approach was trailblazing. She chose a female designer, the gifted Taki Handa, to create the site and employed the Japanese gardeners Jiju Suzuki and Shinzaburo Matsuo to develop and maintain it, aided by estate workers. The UK's first and only Japanese garden of such size and scale to be designed by a woman, this enchanting creation at the foot of the Ochils forms a unique and authentic bridge between Scottish and Japanese culture.

Even as she neared the end of her life, Ella seemed inexhaustible. She supplied clothing for hospitals, socks for sailors and mittens for airmen through her Red-Cross War work, gave talks in London and Scotland as Vice President of the Scottish Geographical Society and, aged 83, took out life membership of the Kinross Beekeepers' Association.

And always in the background was *Sharaku-En*, to which she would retire to read and think in the shade of her 'gazebo'. The garden represented a place of ineffable peace, not just for Ella, but for the many she invited to Cowden to experience its healing powers, including soldiers from both World Wars.

Despite all her adventures, Ella was never happier than when she was at home, wandering around 'a little larger than life in her homespun garden coat lined with the tummies of her own grey squirrels and an amorphous felt [hat] that had known the suns of Samarkand as well as the snow of Scotland'. As an aunt, great-aunt and great-great-aunt, she has

remained a shining inspiration for later generations of her family and heirs at Cowden.

When I married into the Stewart clan 35 years ago, I was totally unaware of Ella and her adventures, but gradually I learned about her and wanted to find out more, realising that we had travelled many of the same routes. I had spent several years of my childhood living in Malaysia, where my parents took us to lots of the places Ella visited. When I journeyed through China in the 1980s, the regulations were being relaxed and I was able to reach Tibet proper. I also went up to Srinagar, and to Leh, where I crossed paths with the Dalai Lama at the Hemis monastery that Ella knew. In the 1980s and 1990s I lived in Japan for five years and also fell in love with Japanese gardens.

Many of my travel experiences, particularly those in China and some of the more distant parts of India, were pretty taxing and I realise now that little had changed in 70 years. Ella may never have had to travel on the roof of an Indian train because there was no room inside as I did, but she was certainly familiar with the kind of accommodation I encountered in truck stops in Xinjiang – a plank for a bed, unspeakable food and equally unspeakable lavatories. Yet, whereas I could have made it to some sort of safety had things turned bad, she had only her own resources to fall back on, which underlines for me how brave and intrepid she was.

The one place I had never been to was Central Asia and so, in 2024 I set off for Ashgabat in Turkmenistan with my daughter Molly. Molly has many

of Ella's character traits, not least a burning desire
to explore the less travelled places. We continued on
through Turkmenistan to Uzbekistan, where we met
up with three other members of the family. Clutching
a copy of Ella's book *Through Khiva to Golden Samarkand*,
we stood where she had stood, compared her photographs with what we were seeing and discussed her
achievements with our various guides. They had never
heard of her and were amazed that they hadn't. They
showed us many old photographs that hinted at how
hard it must have been for a western woman to travel
solo through this tough region in the early 1900s.

I have loved reading Ella's letters (despite her dreadful
handwriting) and sifting through all the photographs of
places we have both been to and things that no longer
exist. Among them were many views of the Japanese
garden she made at her home in Scotland, although
when I first got to know Cowden, I found it difficult to
imagine that the pond fringed by rhododendrons was
once the centre of that beautiful creation captured in
the old pictures. After many years of hard work, overseen by my sister-in-law Sara Stewart, *Sharaku-En* is once
again as it was. Sara's determination and drive has been
phenomenal and she has appeared, at least on the sur-
face, undaunted by obstacles that would have stopped
a lesser person. That it exists at all is testament to the
extraordinary energy, intelligence and curiosity of Ella
that prompted me to research and write about her life
and to follow in her footsteps to Khiva.

LUCY STEWART
Warwickshire · March 2026

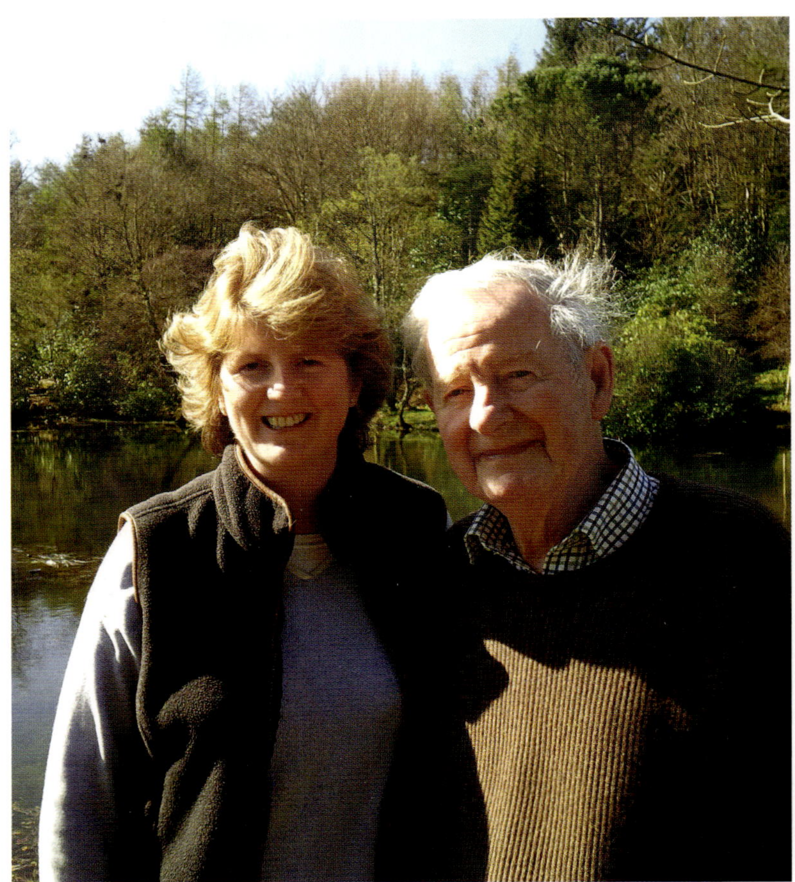

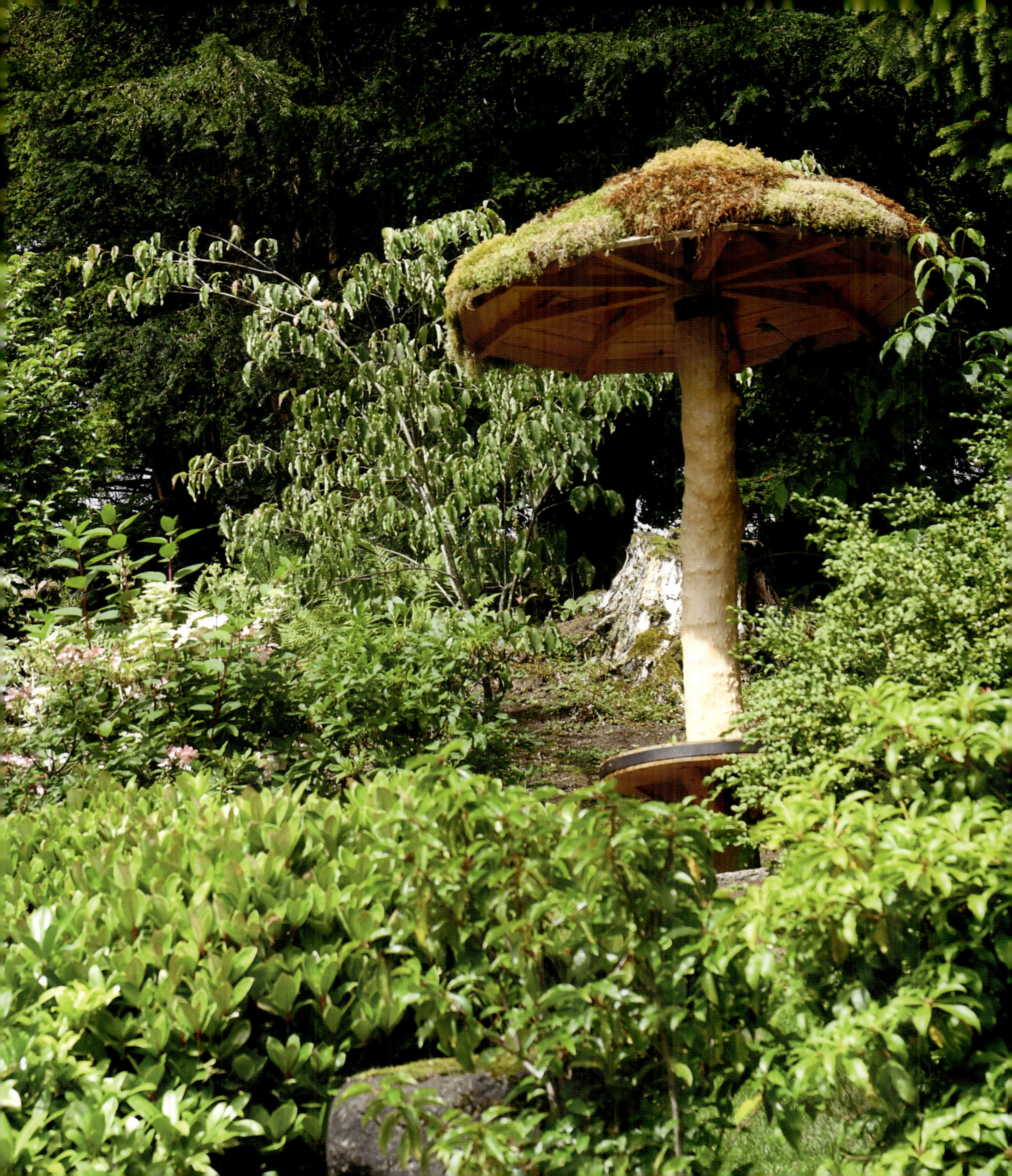

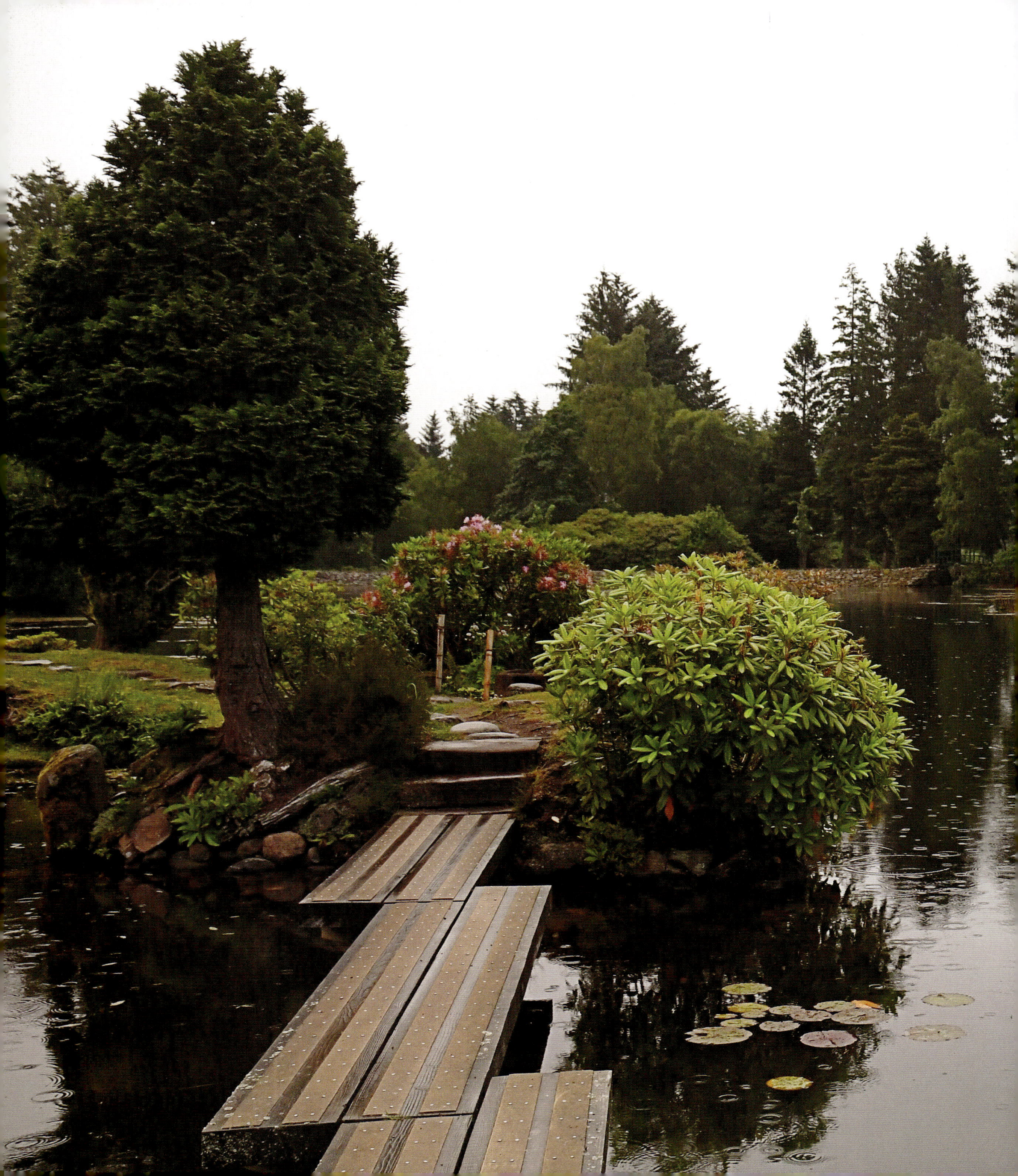

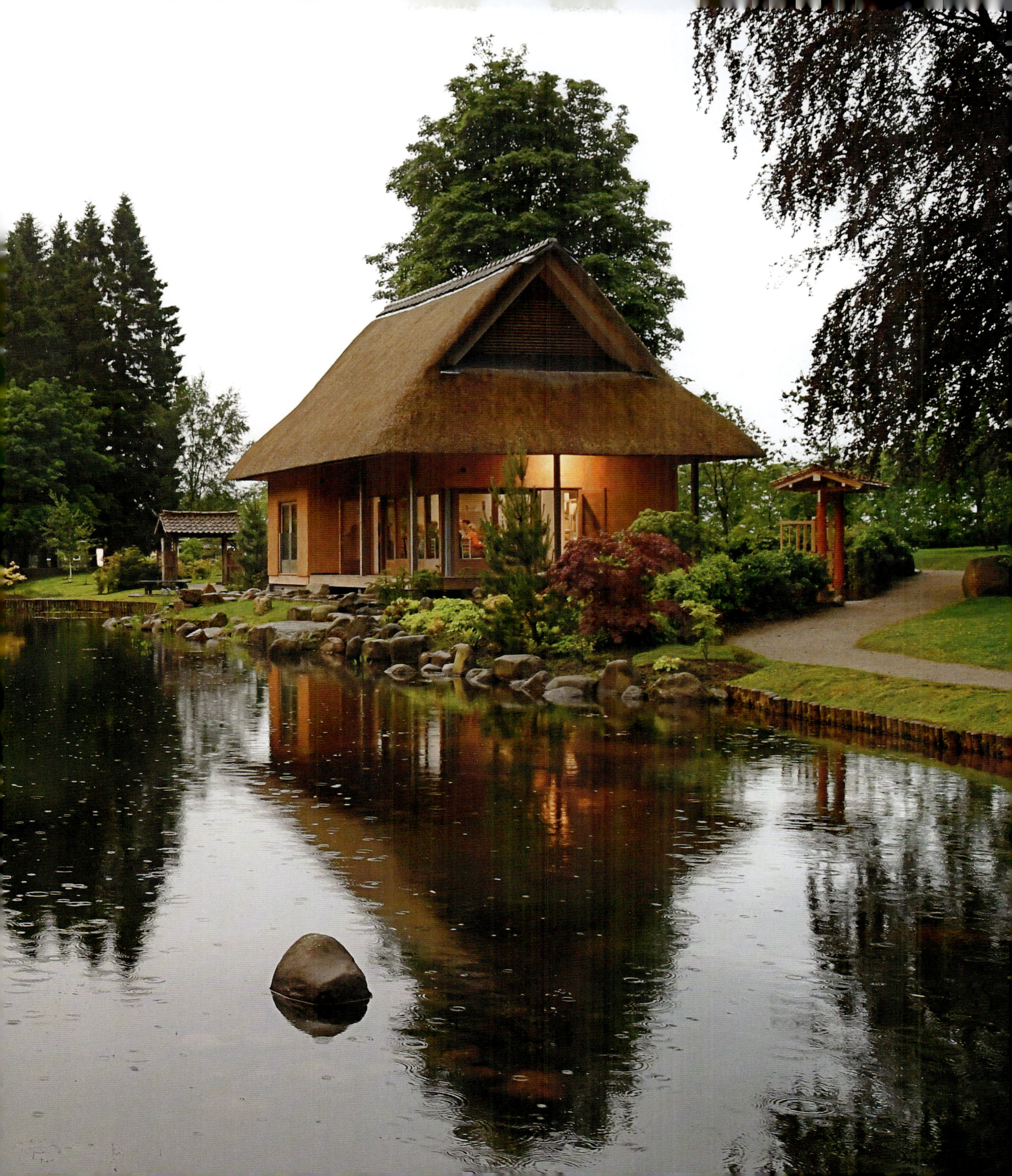

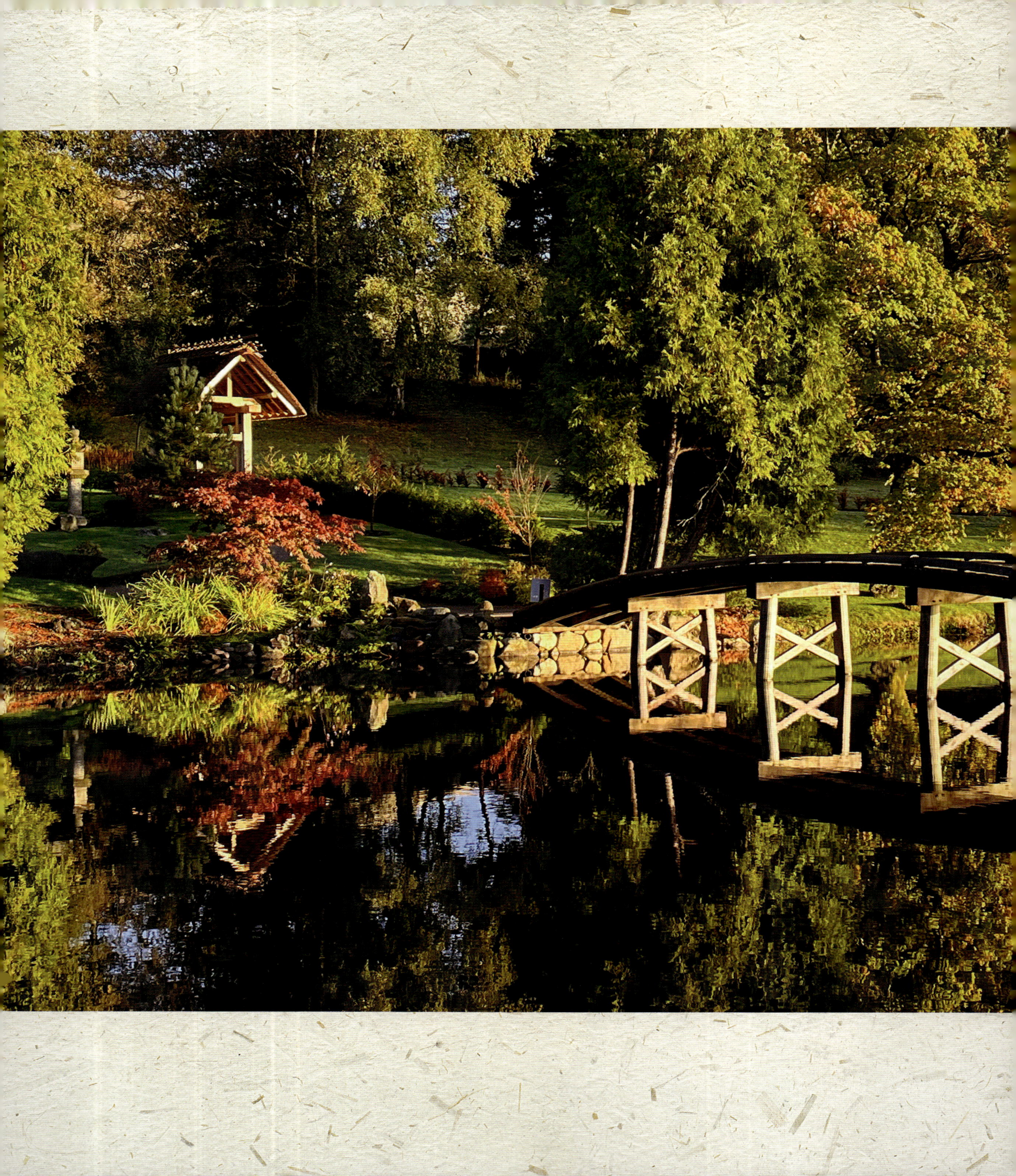

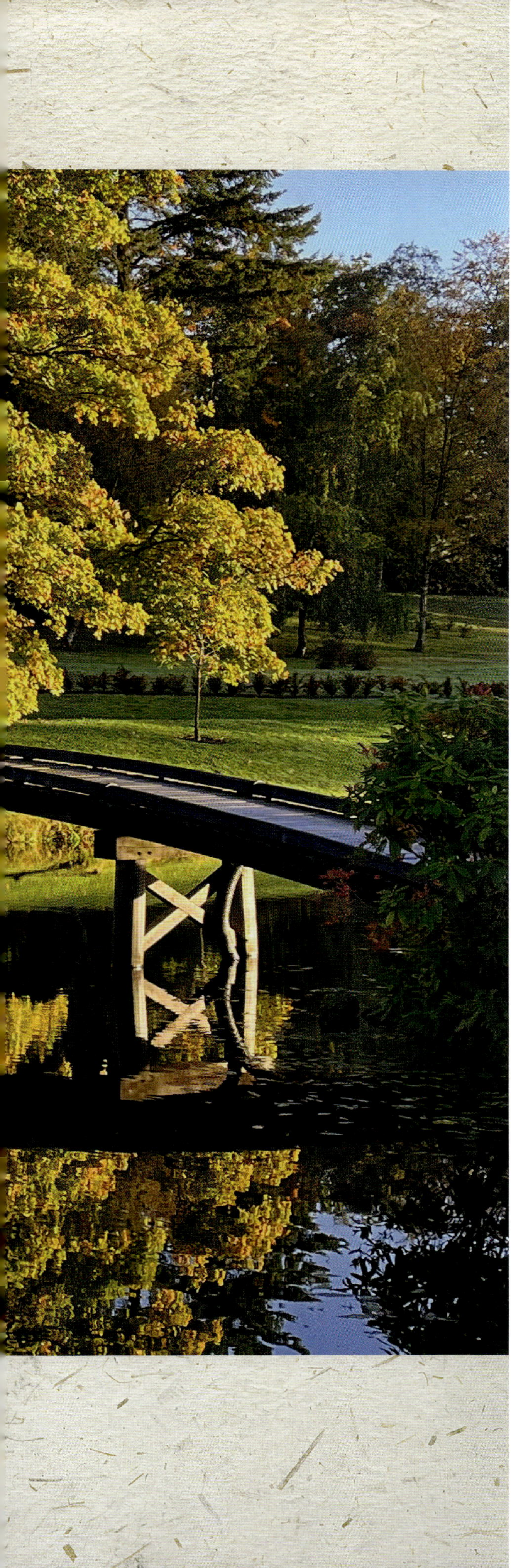

1 · Childhood – the Christies at Home and Abroad

ISABELLA ROBERTSON CHRISTIE, always known as Ella, was born in the parish of Cockpen, Midlothian on April 21, 1861, the second child of John Christie, a mine and iron works owner, and his wife, Alison Philp. John (b. 1824) had inherited a considerable fortune, along with land and property in Midlothian, Ayrshire and Lanarkshire, from his father, Alexander Christie (d. 1859). Alison (b. 1818) was the daughter of William Philp of Dalkeith, who died possibly before she was born, and his wife Margaret Coldwells. Soon after she was widowed, Margaret moved with her infant daughter to live with her bachelor brother John Coldwells at Stobsmills House in Midlothian. Coldwells, who ran Scotland's earliest gunpowder works in Gorebridge, had never liked his brother-in-law and happily accepted his sister into his home, where she acted as housekeeper. He became very fond of his niece Alison, and later of his great nieces, Ella and Alice, to whom he left a significant fortune in trust on his death in 1864.

When Margaret died in 1851, Alison took over running the Stobsmills household. For ten years, she dutifully refused John Christie's repeated offers of marriage until, aged 36, she finally accepted and

Coldwells had to resign himself to losing his much-loved niece and companion to the man who had waited for her so patiently. After their marriage on April 27, 1859, John and Alison lived chiefly at Millbank in Midlothian and produced three children in quick succession: John (born 1860), Ella (1861) and Alice (1863).

In 1865, under the terms of his marriage settlement, Christie was able to use money inherited by Alison from her uncle John Coldwells to buy the 524-acre Castleton Estate in Clackmannanshire. Situated close to the old drovers' village of Muckhart, it lies two miles from Dollar at the southern end of Glendevon, the pass through the Ochil Hills.

According to the historian Ebenezer Henderson LLD, whom Christie consulted for information about his new property, the first known reference to Muckhart is in an early 13th-century charter from Dunfermline Abbey, referring to a clergyman called Muceard. In 1319, William Lamberton, Archbishop of

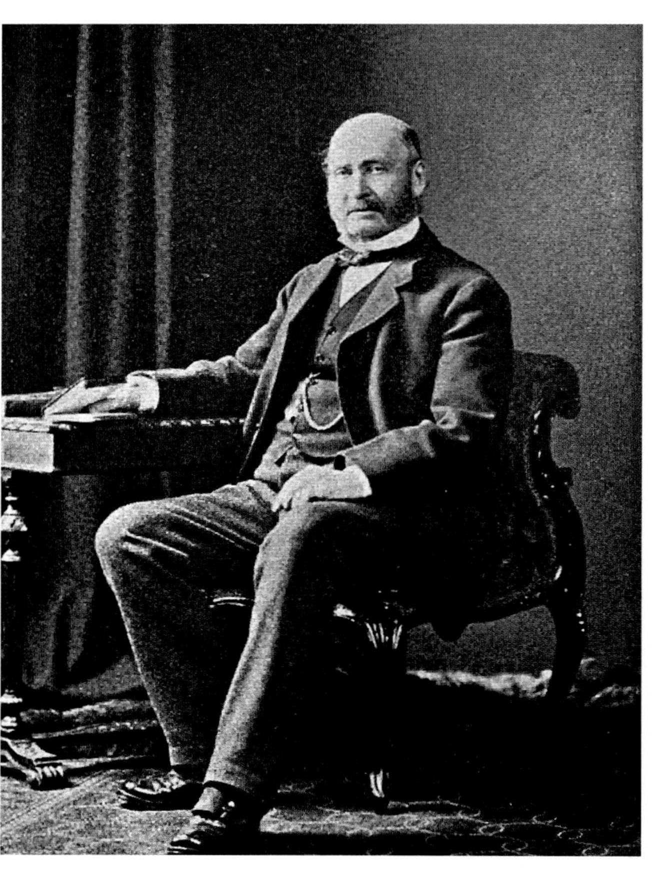

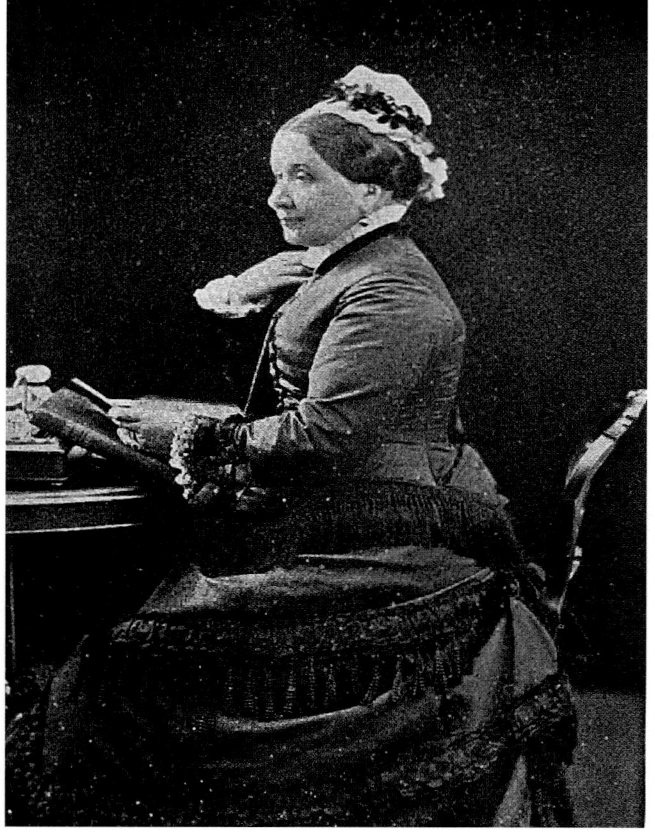

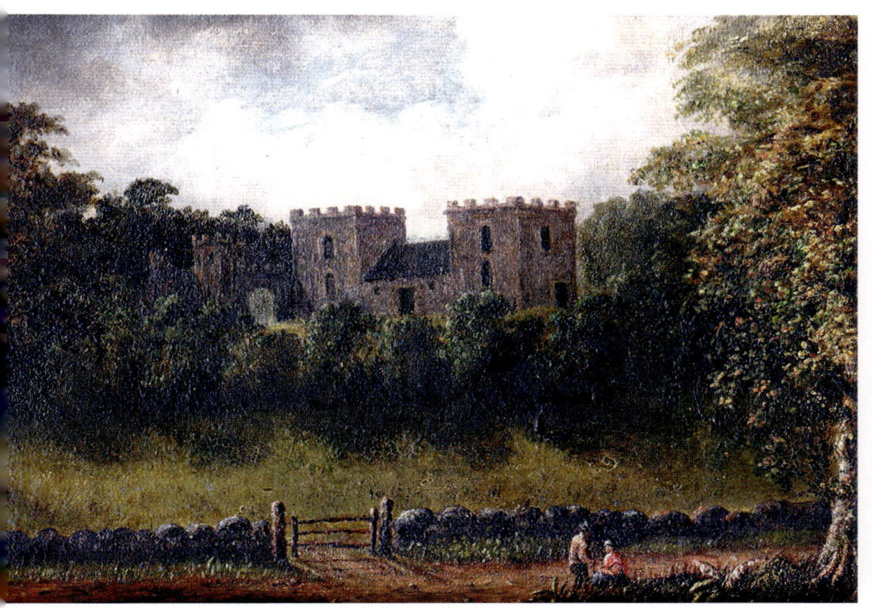

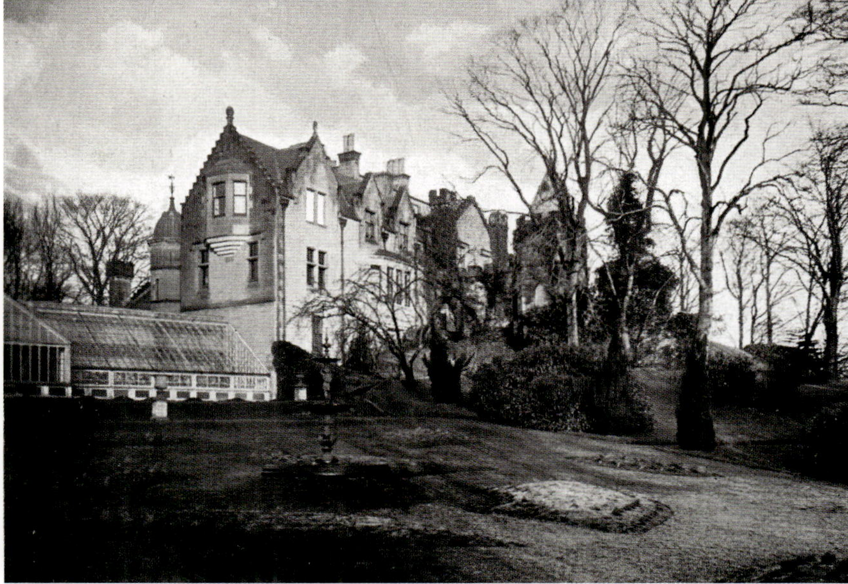

St Andrews, built a "castrum" in "Muckhard Schyse" – 'undoubtedly that castle of which a fragment still remains in the SW [south-west] of your residence,' Henderson wrote to Christie in 1875.[1] This fortified bishop's palace passed to the Earl of Argyll in 1491 and was still defendable in 1645, when it is said to have been one of only two buildings in the parish that were not razed by Montrose's troops, who set fire to the Muckhart church.

Christie renamed the estate Cowden, which met with disapproval from Dr Henderson: 'I may also note that it appears to me that the old name "Castleton" should be retained in preference to the very common name "Cowden". Castleton, of course, takes its name from the name of the castle, which in early times stood adjacent to your residence and of which a fragment remains. Cowden: this name is a compound word derived from the Celtic, which, in its pure, original form, is "Cul-den". From this form, the word, by being mixed up with Scotch, has suffered much, for it has been mutilated into equivalents such as "Col-den" and, in its worst aspect, "Cow-den". "Cul" signifies a back of any place (in Celtic) and "dun" is a hill in the same language – hence "Culden" simply means "the back of a hill" – the back of any hill'.[2]

From 1867, Cowden was Christie's principal family home. One of the reasons that he bought the estate rather than staying in Midlothian, where the soil is better and the rainfall less, was his interest in growing plants and trees that thrive in acid soil. He filled the grounds with assorted rhododendrons, which were greatly in vogue at the time, and with a considerable collection of specimen trees and conifers, which he hoped to see to maturity during his lifetime.

As appropriate for the country seat of a gentleman, he had the house remodelled. This work was carried out in 1893–94 by Honeyman and Keppie and included an octagonal tower with a wrought-iron weather-vane – an early design by Charles Rennie Mackintosh, who was employed by the practice at the time. Christie also filled Cowden with an eclectic mix of furnishings and *objets d'art,* including ebony cabinets from Paris, carpets from Smyrna, Venetian chandeliers, collections of shells, paintings by successful artists of the day, Meissen china, alabaster urns and Dutch-tiled chimneypieces. He had a great interest in ceramics and

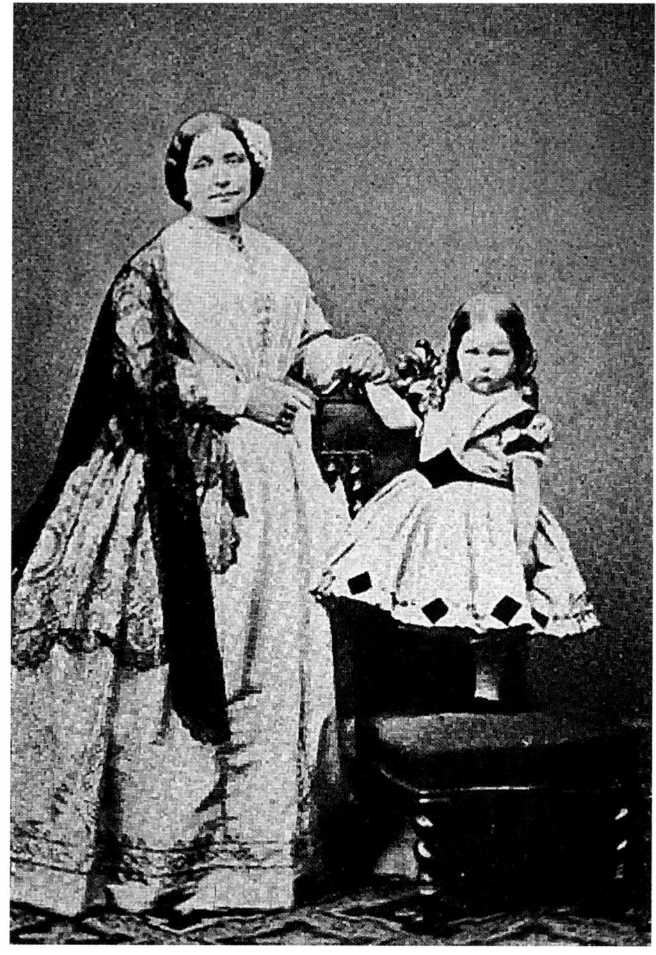

many of the trips he made with his family to Europe
were organised around visits to manufacturers so that
he could build up his collection.

Cowden was equipped with all the furnishings,
objects and appliances that underpinned the family
life and domestic operation of a Victorian country
house. One bathroom had a mahogany lined zinc bath
raised on steps that took the housemaids an hour to
clean with bath-brick and soda. There were embroi-
dered Parisian sheets stored by the laundress in the
linen room, bundles of quills and reams of note paper
in the stationary store, plain nursery food served on
cream-coloured Wedgwood china decorated with the
word 'Nursery' in brown script, and a wide 'pram'
resembling a miniature bath-chair, with one small
front and two larger rear iron wheels, a waterproof
apron instead of a hood, a stuffed and tufted back and
a seat covered in crimson Morocco leather.

John and Alison Christie were both well-educated
and, naturally, they wanted the same for their children. They decided not to put them through formal
schooling but to educate them at home with the help
of governesses. They engaged a Swiss nurserymaid,
Susanne Kuenzi, whose main task was to ensure that
the children were fluent in French, and also ensured
that they acquired a working knowledge of Italian,
Spanish and German. Ella would later add Swedish
to the list and translate a volume of Finnish fairy tales
by Zach Topelius, her first published book. The Christies were also musical – Alison played the piano by ear
and John loved Italian operas. Ella was taught to play
the harmonium and would later often be called upon

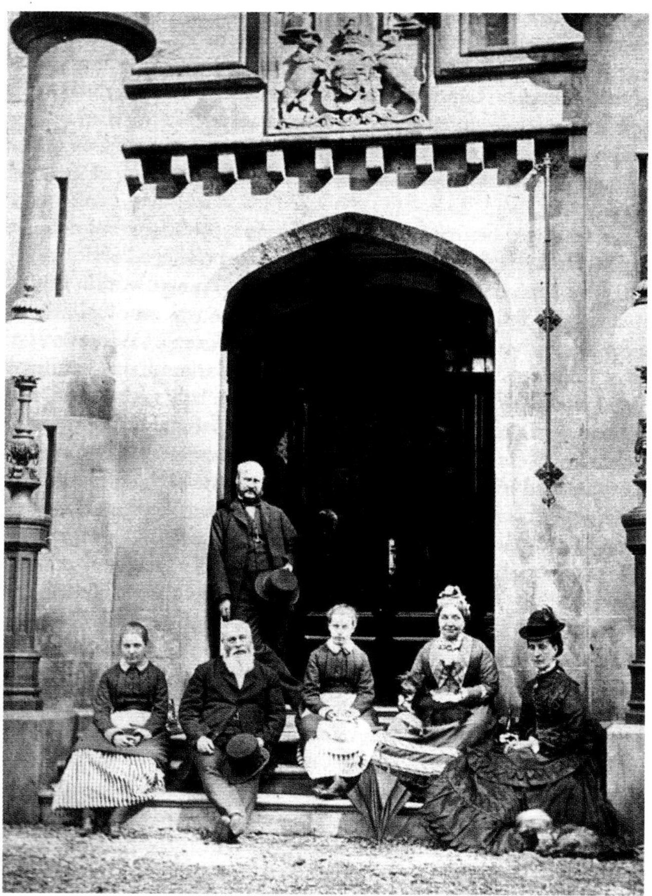

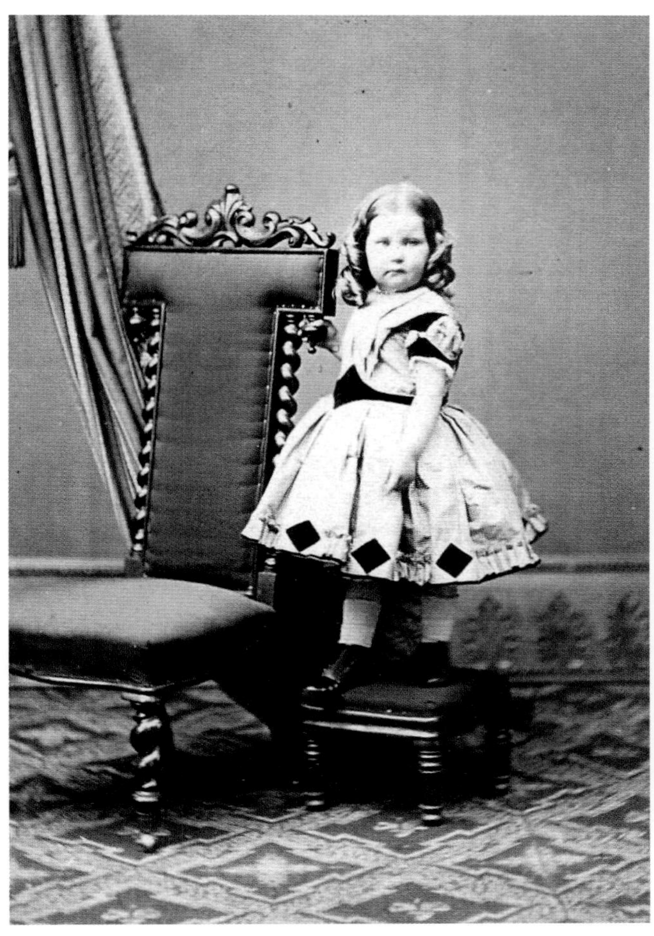

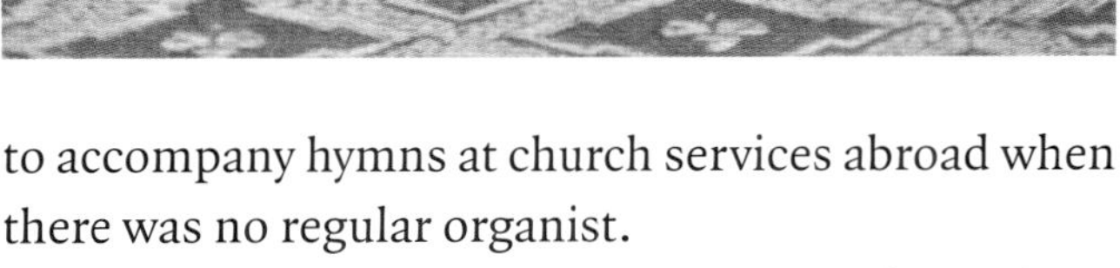

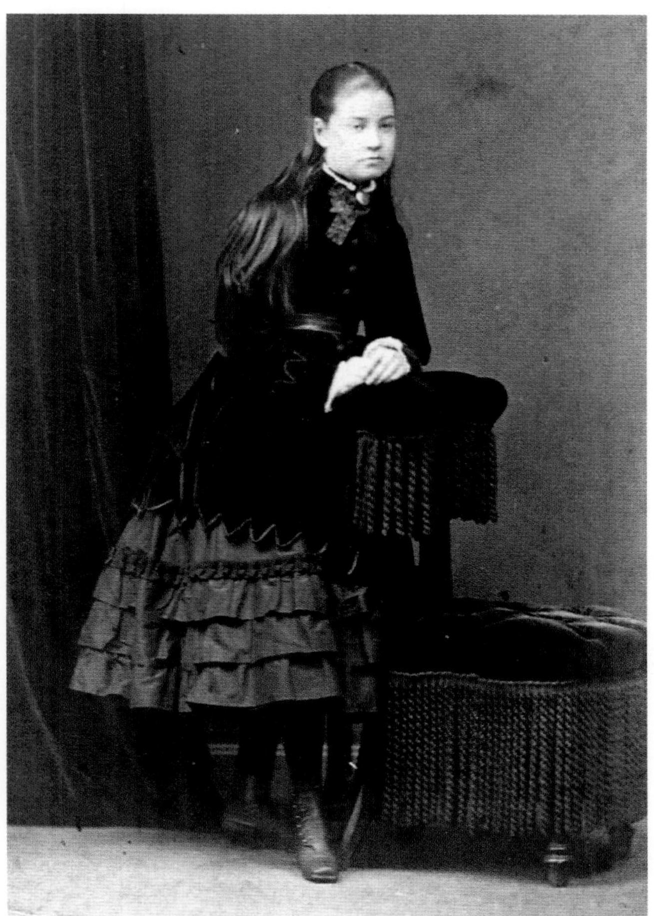

to accompany hymns at church services abroad when there was no regular organist.

The family spent the summer at Cowden, where activities revolved around ponies, dogs and picnics. The sisters' governess, Jane Semple, taught them 'the love of all country things. There were scrambling walks … brambling excursions, gipsy teas, making toffee on the schoolroom fire, and assisting in our gardens – all were included in Miss Semple's curriculum'.[3] The gardener James Henderson taught the children the rudiments of grafting, propagation and cross pollinating to produce new varieties. Indoors, there was lots of reading and needlecraft. The village shop offered a tempting array of sweets, including a mix of 'rosebuds' and acid-drops in a wide-mouthed glass bottle, 'conversation lozenges' and – one of their favourites – a treacle confection called Gundy. Ella's greatest treat, however, was tea at the baker's, when the pastrycook's wife, in black, lace-trimmed cap, lavished them with cookies and shortbread.[4]

Growing up in this privileged world, Ella and Alice enjoyed a very happy and comfortable childhood. They spent little time with other children, possibly in part because their rather older parents, who were both only children, treated them more as mini adults. However, they had plenty of attention from the adults of the household, in particular the head nursery maid Isabella Thorburn, known as Baa, who was 'austere and gloomy,' but 'a miracle of patience and kindness throughout our childhood'.[5]

Winters were spent in Edinburgh, firstly at Brae Lodge and then, from 1874, at 19 Buckingham Terrace. It took all day to get there from Cowden. They would go by horse and carriage to Dollar and then take the single line train to Stirling, where they changed onto the main line to Edinburgh, travelling the final leg by carriage. The domestic routines continued much the same as at Cowden, with a few variations. The sisters attended deportment classes at Miss Oliver's Academy at 10 Rutland Square, which gave Ella the great bearing that she found so useful when faced with officialdom in far flung places.

The Christies had a Victorian commitment to charitable works and Alison would organise bazaars in the names of her children, who then chose the charities to which the proceeds were sent. Religion played a dominant role in their lives. On the Sabbath, the children would go through the catechism in their father's study and attend two Presbyterian church services, so that over time they got to experience the various different preaching styles popular at the time.

Their older brother John was a sickly and mentally handicapped child whose death, in 1872 aged 12, plunged the household into full Victorian-style mourning. 'For our brother, when we were nine and eleven, we had black merino frocks trimmed with bands of crêpe, like little widows. Six months later, our summer hats, bought in Paris, were black chip adorned with black crêpe flowers, and eighteen months later, when for the following summer our mother suggested white chip with black velvet bands, Baa held up her hands and said she "did not know what the world was coming to"'.[6]

John Christie believed in broadening his children's minds with experiences such as taking them down a coal mine and visiting various industrial works. Travel was also an essential part of their education and, from 1873, the family would make an annual trip to the Continent, including Spain, Italy, Germany and the Low Countries. They would travel to London by train – in those days, a far from luxurious experience, with unheated carriages, no lavatories or refreshment car and a single oil lamp in a niche over the door providing each compartment's only light.

On their first trip, in the spring of 1873, the girls were enthralled by London, with its horse-drawn buses with straw underfoot and the conductor hanging from a strap on an external ladder shouting 'Penny all the way!'. They were captivated by the shops and arcades selling all manner of enticing, colourful and ornamental goods, and by the waiters and bustle at the Euston Hotel, but they hated the novel experience of a ride on the new-fangled underground railway, as 'the suffocating atmosphere of the sulphurous fumes made us choke and cough for a considerable time'.[7]

After taking a train to Folkestone and a paddle steamer to Boulogne, the family continued to Paris, where they visited the Louvre, warned by Alison not to mention Napoleon. 'The Tuileries were then a blackened mass of ruin, so was the Hôtel de Ville, the Place Vendôme was graced only by the pedestal of its column, and the palace of Saint-Cloud was a pathetic ruin'.[8] But they walked a great deal in the parks and did a lot of shopping, buying clothes at Bonmarche, a department store that the sisters would return to throughout their lives. Then on to Cannes, again by train (again very basic), where they spent two months among terraces of olives and vines in the Hôtel d'Hollande. The winter sunshine, the picnics on La Croisette, the market places, even the church services followed by a stroll round the local cemetery as a suitable form of sightseeing for the Sabbath – all these experiences made wintering on the French Riviera infinitely preferable to remaining in freezing Scotland. In 1876, they saw the Paris Exhibition, with its Eiffel Tower and Trocadero Palace, visited Florence and returned via Belgium and 'its famed battlefield, where mamma described to us all she remembered of what Sergeant Cotton, many years before, had told her on the ground where he had actually fought'.[9]

On their last family holiday, in the winter of 1879, Alice met Robert King Stewart in Paris and they fell in love. She was only 16, but the couple had an understanding and, as soon as she turned 17, they announced their engagement. They were married at Cowden in October 1881 and Alice went to live at Murdostoun Castle in Lanarkshire, where she embarked on a life of dedication to the estate and the wider community. Her father-in-law, also Robert Stewart, had been Lord Provost of Glasgow and was responsible for the great scheme that supplied the city with safe water piped from Loch Katrine, opened by Queen Victoria in 1859. In the same pioneering spirit, his eldest son Robert installed electricity at Murdostoun as part of renovations carried out by the architect John Honeyman; the house was one of the first to blaze with electric light when the power was turned on in September 1882.

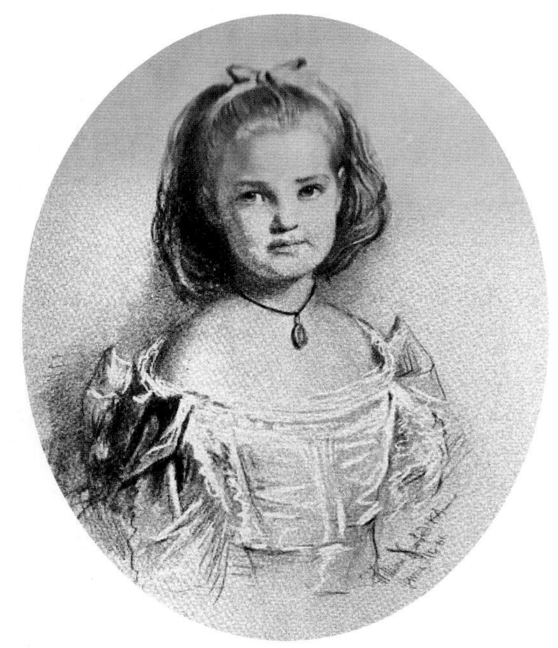

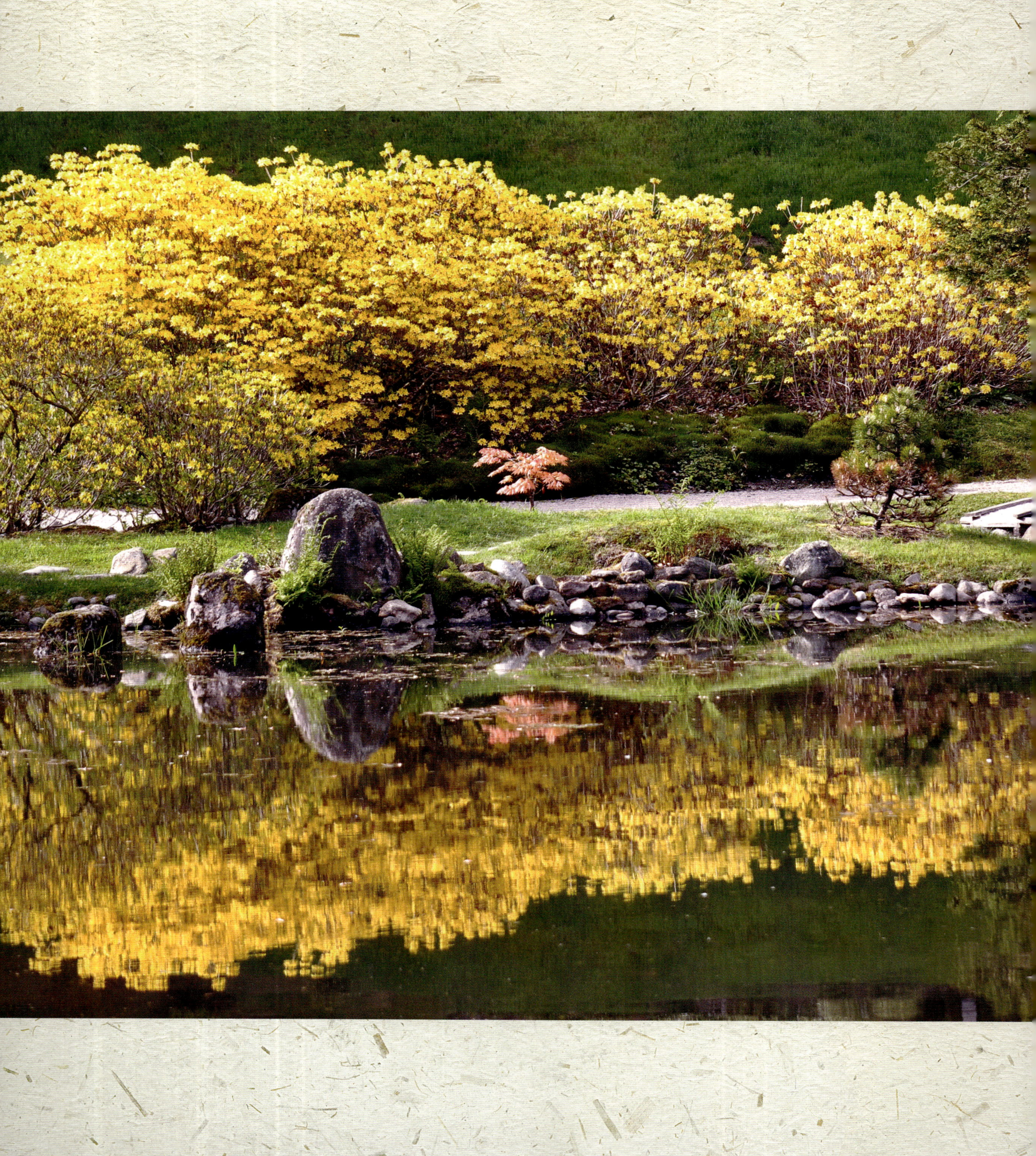

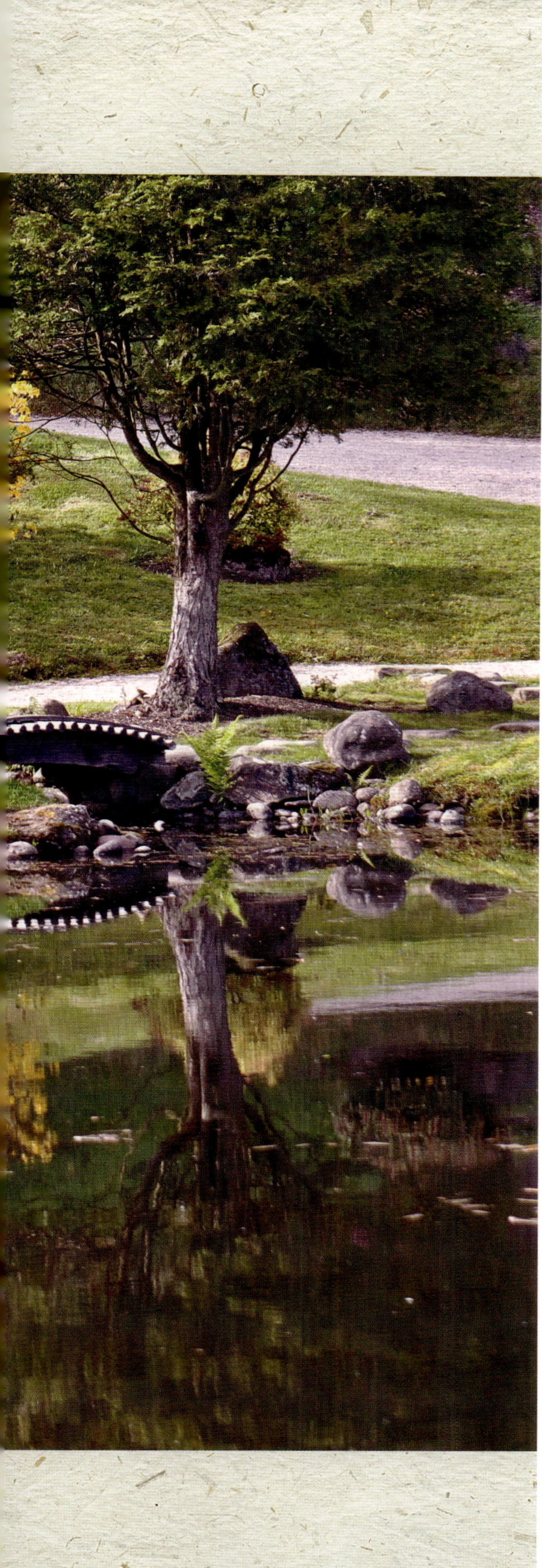

2 · Young Adulthood – Trials and Pleasures

AFTER ALICE'S MARRIAGE, the two sisters corresponded on an almost daily basis for the rest of their lives. Their letters, which survive in the family archive and the National Library of Scotland, are full of detailed accounts of their experiences, from Alice's accounts of life in Lanarkshire – race meetings, hare coursing, a week-long International Aviation Meeting held at Lanark in August 1910, her work with the Queen's Institute of District Nursing and the British Red Cross Society – to the challenges of running a big house and, of course, Ella's travels. There were frequent visits between the two estates and Ella became very fond of her three nephews, writing affectionate letters to them and mentioning things that she thought they might find interesting when she wrote to Alice from far flung places.

Alison Christie had become less robust, so Ella became their father's travelling companion. Her letters to Alice contain snapshots of their wanderings: 'they make an excellent Omlette Surprise at the Jules Cesar at Arles, be sure to order one if you go there … We got a good plate of majolica in Granada, but the curio shops are poor… I found [in Norway] a delightful recipe for rhubarb, you boil it till tender and put it

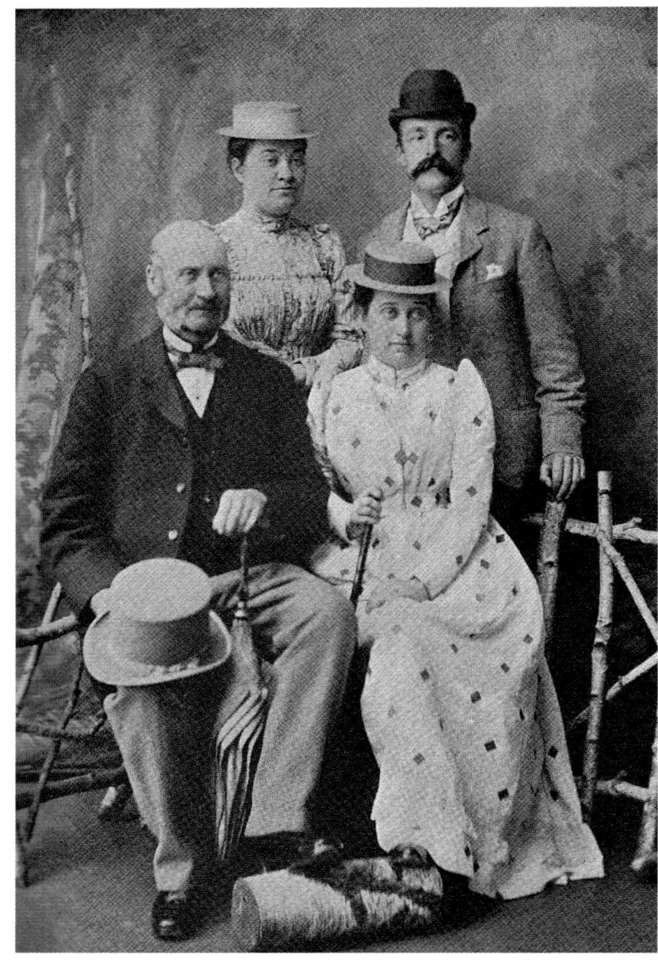

through a cloth with fine sieved almonds … I have sent half a dozen tins of *foie gras* home from Toulouse.'[10]

In the spring of 1887, John Christie suffered an attack of pernicious anaemia, a blood disorder that left him seriously ill and unlikely to recover. He slipped in and out of a coma, with Ella sitting for hours next to his bed feeding him small sips of brandy – he refused to be nursed by anyone else. Eventually, however, he regained consciousness.

It took Christie eight months to recover, but he was very changed from the affectionate father and husband his family had known. He seemed to have lost all love for Alison, turned against Ella and was indifferent to Alice. He became very secretive and would lock all the doors in the house, particularly those to his dressing room. He shunned company and became angry if Ella invited people to Cowden without his prior agreement. He would disappear off to Edinburgh, telling nobody and carrying a big, black, heavy bag, the key to which he kept in his coat pocket. One day in the summer of 1889, he came downstairs and announced that he was leaving for the Continent with a Dr Cunningham, whom none of the family knew. He was away for six weeks.

In the spring of 1893, Christie had another attack and sank into a depression. Ella, Alice and Robert decided to take him to Sweden and Norway in an attempt to cheer him up. The trip was punctuated by episodes of difficult and strange behaviour. When a telegram arrived with the news that Alison was dangerously ill, her husband appeared totally unmoved. Ella and Alice rushed home, leaving Robert to accompany

Christie back more slowly. After a six-week illness, their mother died at Cowden on October 13, 1894. Christie seemed barely to notice his wife's death and his behaviour left his daughters profoundly shocked. He refused to co-operate over the funeral arrangements, disputed bills paid with Alison's money and ploughed up a stand of ornamental trees that she had particularly loved.

Alice and Robert would later testify to many other character changes – the secrecy, the shunning of friends, the apparent dislike and increasingly hurtful treatment of his once favourite Ella, who now had to play the role of dutiful daughter 'to a secretive and highly eccentric old tyrant'.[11]

Despite all this, in February 1895, Ella took him to Cairo with her maid Prudence Humphries and a Miss Hall, who seems to have been the friend she refers to as May in her letter-journal written in the Howard's Hotel, Jerusalem on March 17: 'May attended a Greek service there [the Church of the Holy Sepulchre] this morning, and had her parasol stolen. I did not go as the smell of those Russian pilgrims is too awful …'.[12] Three weeks after their arrival in Cairo, Christie had decided he didn't want to stay and made a sudden departure with the tour guide, leaving the ladies to continue without him. After travelling to Jerusalem, they explored Palestine with a dragoman and Turkish escort and then got a Russian steamer from Jaffa to Beirut, from where they continued on to Damascus. Ella was fascinated by the city, with its surrounding orchards just coming into blossom, its bazaars with 'wide vaulted arcades, and such goods I never saw before, embroideries and carpets. I bought some of the former, and a robe of white silk on which is written all over in stripes "May Allah give you good health". The caravan from Bagdad has just arrived, hundreds of camels, I never saw such a sight'.[13] By April, she was back in Scotland.

Eight years later, on August 19, 1902, Christie finally died at his house in Edinburgh, his daughters by his bedside. The degree to which his character had changed was confirmed the following morning, when the doorbell rang and an unknown man walked in. 'I have come to tell you that your father has left everything away from you,' the stranger announced. Out of shock, the Christies let him in and he proceeded to go round the house making an inventory of all the contents. The only thing that escaped his notice was a silver teapot that had belonged to their great-uncle John Coldwells, which the old cook, Rachel Adamson, had hidden under her apron. She later gave it back to the sisters, saying 'Jist tak' it and sae naething aboot it!'.[14]

For Ella, in particular, this was a financial disaster. All she had now was the teapot and a small income inherited from Coldwells. Alice, having married into a wealthy family, was more safely provided for. After much soul searching, and in spite of all their friends and the family lawyers advising them that there was no point in contesting the will, they decided they could not let their inheritance go without a fight, and so they engaged their lawyers, W. & F. Haldane of Edinburgh.

The case came to court in the summer of 1903 and took a week to be heard, keenly followed by the Edinburgh public. Ella would later recall coming out of the law courts and seeing the headline CHRISTIE WILL CASE SETTLEMENT blasted across the newspaper posters, with the death of the Pope announced in smaller letters beneath. Clairvoyancy and a belief in Second Sight was very fashionable at the time and Alice went to see Madame Van Bien, who advertised on the front page of the *Scotsman*. The well-known mystic reassured her that the outcome would be good. Towards the end of the week, Ella was informed that her opponents were in tears and would like to settle the case with a compromise. Painful as it would be for the Christies to have some of their evidence aired in public, Ella's response was 'I don't care if they are in hysterics, we're going on'.[15]

It emerged that Christie had made a series of wills leaving the vast majority of his fortune to his Christie Homes for Orphans, which he had established without telling his family. In court documents relating to the case, his lawyers confirmed that he had conceived the organisation in 1887, but that it was not until 1892 that he had bought the first home, Inverey House in Portobello, Edinburgh. This was followed by a second home, Tenterfield in Haddington, East Lothian, acquired in 1897. The lawyers acting for the trustees of the homes asserted that it was modesty that had stopped him from telling his family and friends about the generous initiative.

Christie had filled Inverey and Tenterfield with silver from Cowden – the explanation for those heavy, locked bags – as well as assorted pieces of furniture and paintings plundered from his house at 19 Buckingham Terrace, Edinburgh. The lawyers argued that, as the object of the homes was to train female orphans for domestic service, he had decided that the girls needed to practice cleaning real silver.

By 1895, just six months after Alison's death, he had had three short interviews with a Miss Sidney Gabbett, with a view to hiring her to run the orphanages. During the third, he proposed marriage to her. A series of witnesses, including the young lady herself, Christie's valet and lawyer, Ella, Alice, Robert and various friends, all testified to the dramatic deterioration of Christie's mental state, or testified that they had found his behaviour bizarre from the start. The

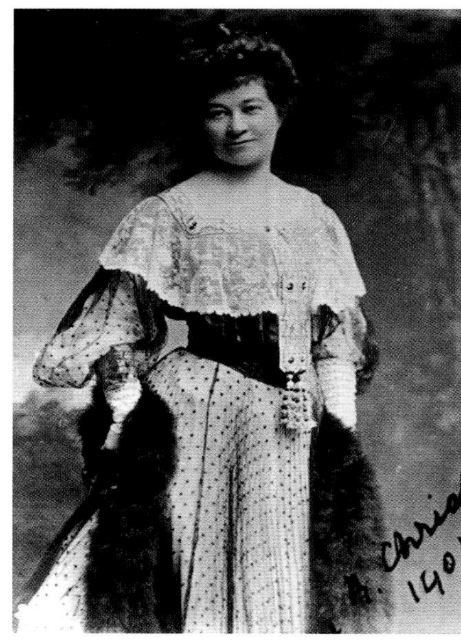

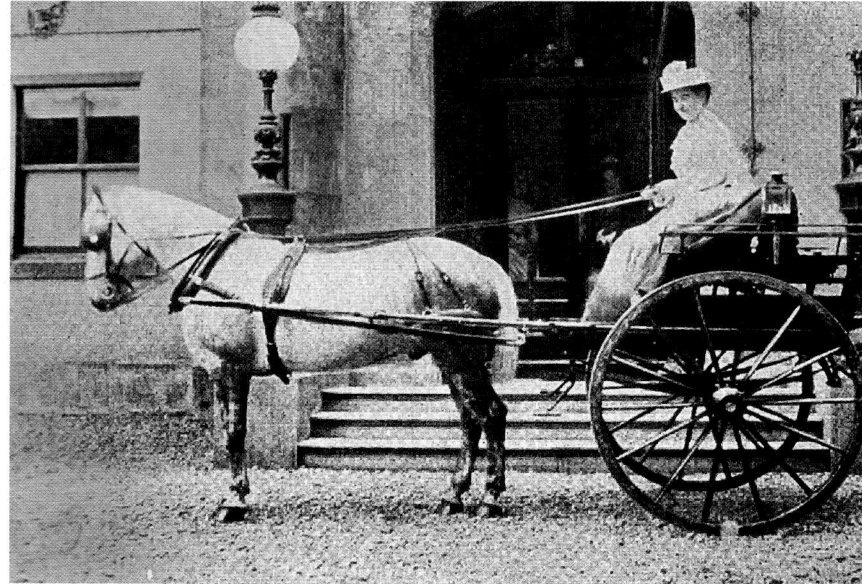

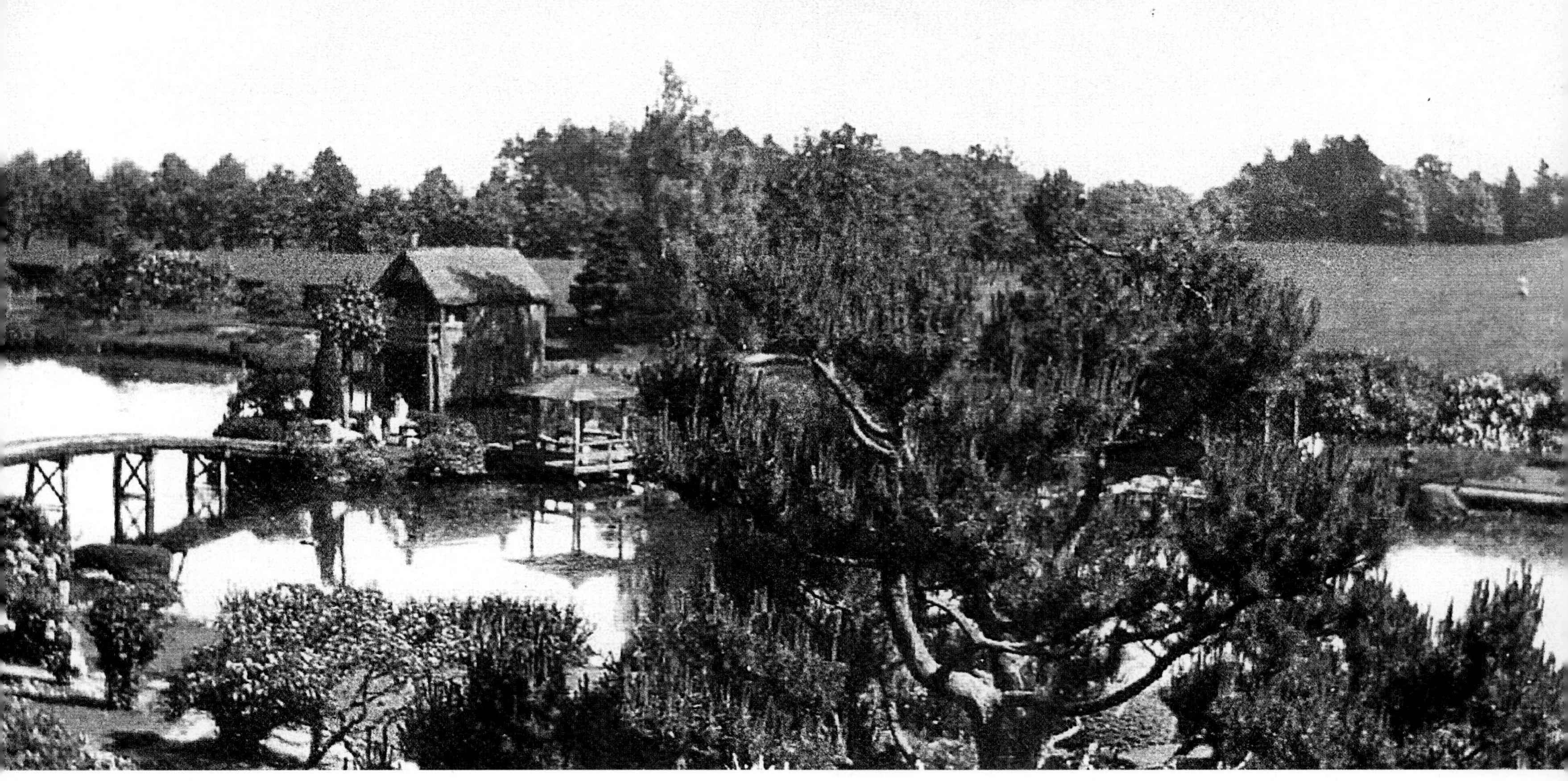

jury found in Ella and Alice's favour, concluding that their father was not of sound mind when he made the series of wills after 1887.

The sisters' strongly developed sense of charity ensured that the orphan homes would get enough money to carry on their work. The remainder of their father's estate was divided between them. Cowden and the Buckingham Terrace house in Edinburgh went to Ella, as intended in Christie's original will. Alice inherited the estates of Milnwood in Lanarkshire and Glenfarg in Perthshire. As Ella's friend Elizabeth Haldane wrote to her on July 21, 1903, 'What a blessing your worries are at an end. You must have had a fair taste of purgatory or something worse… They say no case has excited the same interest and the feeling seems all with you. You are associated with so many functions in which you have helped that you have a large acquaintanceship with every class. Well now, Ella, I picture you as the chatelaine of Cowden and beginning to spread sweetness and light all round'.

Ella took over running the estate with a firm but fair hand. She dealt with her tenants directly and personally oversaw all aspects of life relating to the house and land. Her butler, Nicholson, who knew as much about the place as she did, was left in charge when Ella went away, and even attended auctions to bid on her behalf so that she could build up her father's ceramics collection. As was common at the time, Cowden was let out to tenants when she went away for a prolonged period. Her Blenheim spaniels would be sent to stay with Alice, but Nicholson remained in charge.

Ella was now deeply immersed in the creation of her Japanese garden, which she completed in 1907. That same year, she had the entrance hall at Cowden remodelled by H. Ramsay Taylor to accommodate a teak staircase that she had commissioned to her own design while in Burma in 1904.[16]

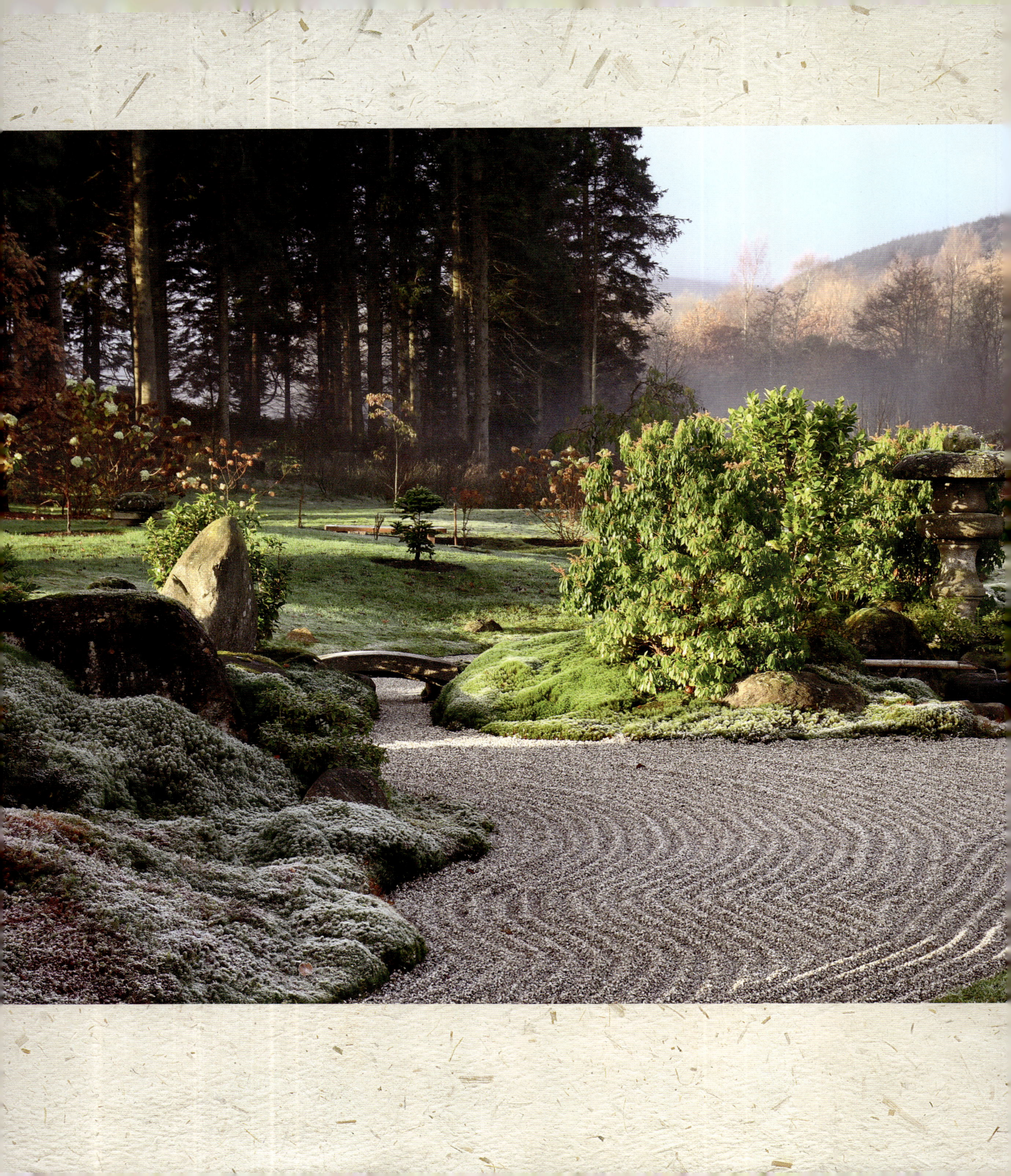

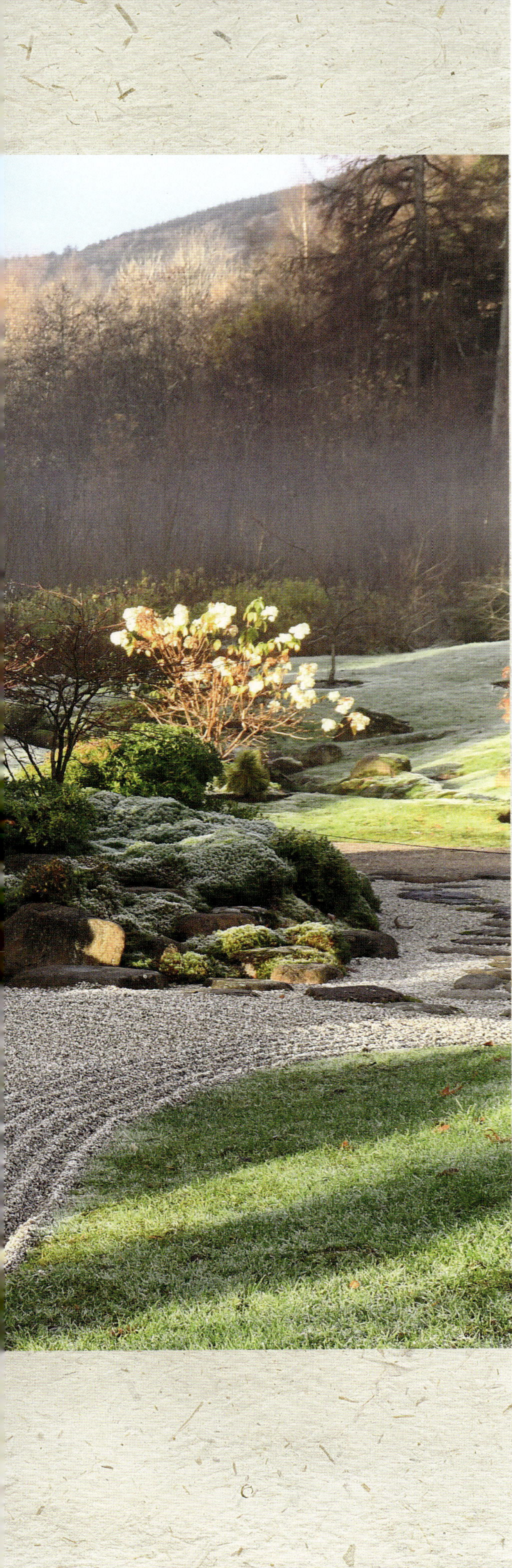

3 · Ella the Traveller

From a child the glamour of the East seemed to cast a spell over my life. Whenever I met a child stranger, my first question always was: "Were you born in India?" If the reply were in the negative all interest ceased for me, as the fact of being born in that land of enchantment seemed to me the most enviable of all to claim as a birthplace.[17]

Miss Christie is one of the great travellers. I do not mean by that merely one who covers immense distances, meeting amazing adventures on the way. There is nothing of the stunt traveller about Miss Christie. She is rather a patient, indefatigable, observant and very thorough seeker after knowledge. She is possessed of invincible good humour and the true philosophic spirit which accepts conditions which would daunt another. Her imperturbable cheerfulness, flexible mind and unlimited powers of observation enable her to get the best and the most out of her amazing journey, such as few women would have had the courage to take.[18]

As a child, Ella loved reading. Particular favourites were *The Swiss Family Robinson*, first published in 1812, and *Leila; or, The Island* by Ann Fraser Tytler (1839) and it is likely that these and other novels inspired her fascination with travel from an early age. Her father fed this enthusiasm with tours around Europe, but it wasn't until after his death that she was free to go where she pleased and for as long as she wanted and her trips became more ambitious. With the great developments in mass tourism, by the later 19th century many people who could afford it were seeking out destinations further afield, albeit still mostly sticking to the beaten track. Ella, on the other hand, was always looking to discover something more interesting around the next corner. She was particularly drawn to Central Asia, firstly by 'the extreme desire to see for myself what lay on that comparatively bare spot on the map east of the Caspian Sea, a stretch of some 1200 miles to the Chinese frontier; and secondly, the lure of those magic names, Bokhara and Samarkand, renowned in history as well as in the pages of classic tales and poetic fiction'.[19]

Closer to home, her interest in exploring places off the beaten track found her 'making a most interesting expedition' to Andorra. She started in Pau, the fashionable French summer resort frequented by Edward VII and his circle, where she had many friends. There she enjoyed all the activities favoured by the leisured classes during the 'gay season,' including attending the races, concerts and the opera, playing golf and

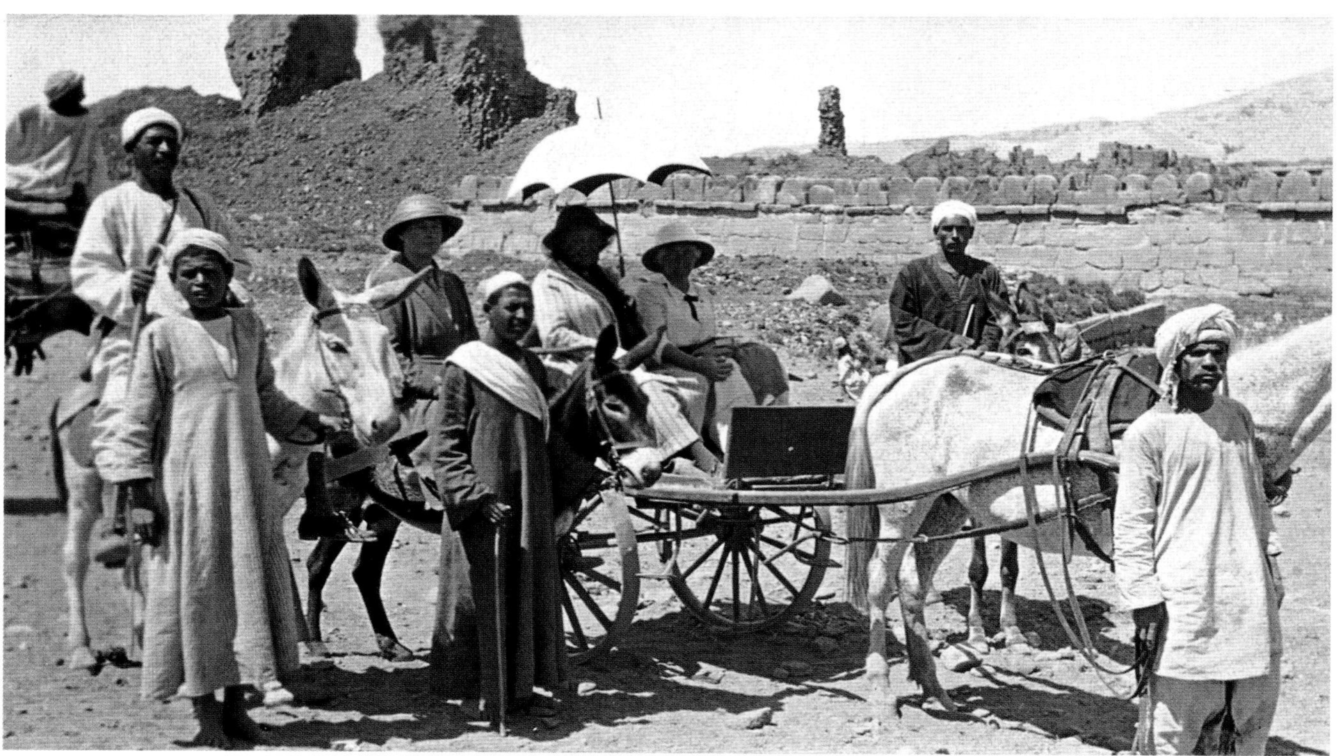

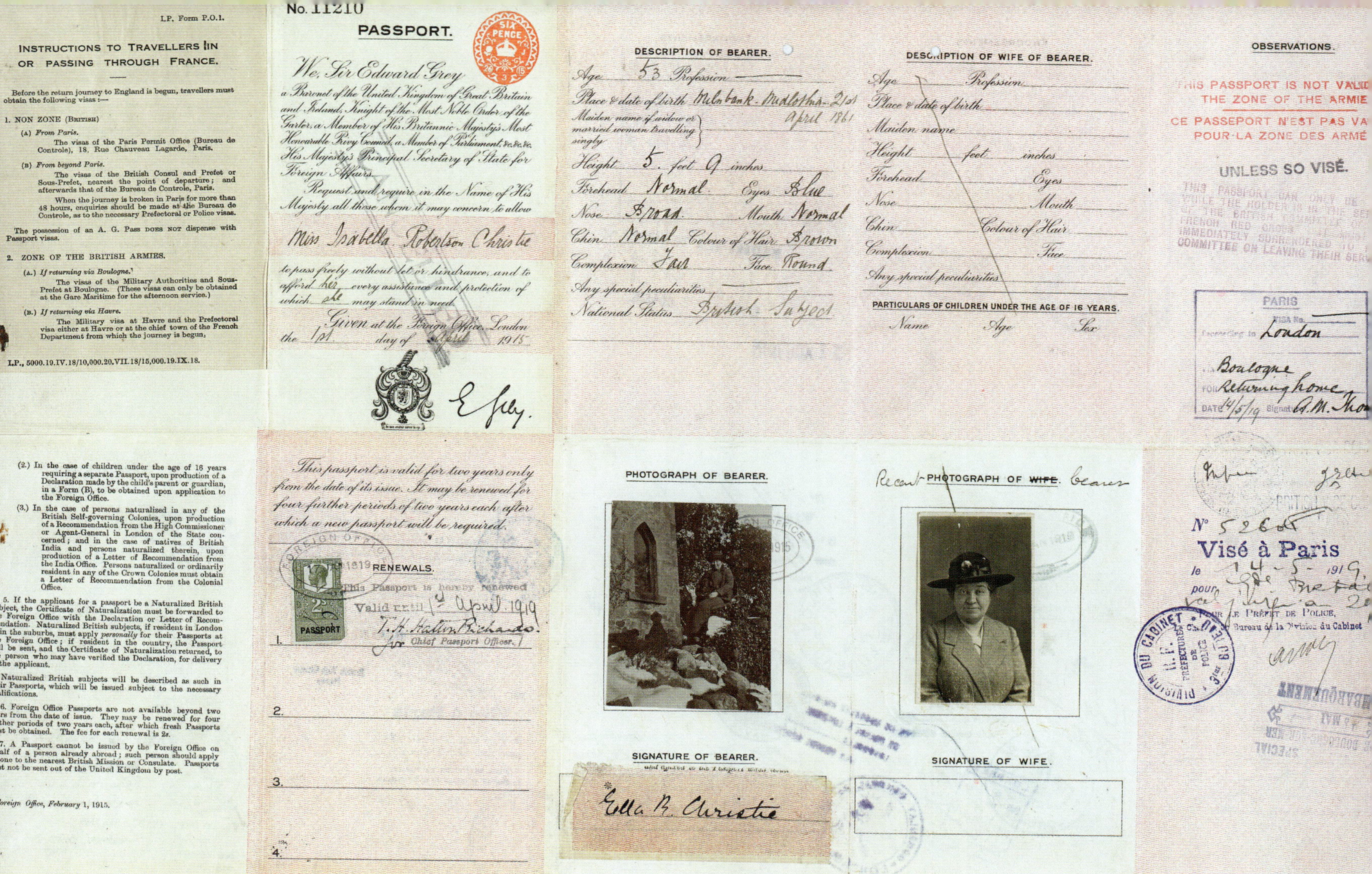

making picnic excursions. She then 'toured along the east end of the Pyrenees via St. Girons and Foix … There left the maid on Monday and pursued my route to the Republic of Andorra'.[20]

There were no roads into Andorra, so Ella travelled on smugglers' paths, taking with her a borrowed revolver and a couple of hired horses and employing among her guides 'the *head* smuggler – and swindler. He charged me 140 francs for four days'.[21] To reach her destination, she had to trek for four hours through deep snow over an 8,339 ft pass, skirt down to her ankles, printed headscarf and stout stick.

Ella was clearly tough, but she was also blessed with a good deal of luck and the ability to make the most of it. Halfway through this journey, she met the president of this ancient little independent republic and charmed him sufficiently in her bad Spanish to elicit a letter of introduction that opened doors everywhere she went.

This combination of stamina, fortitude and luck would stand Ella in good stead on many future trips. She would make five more epic journeys, which are recorded in her long letter-journals to Alice and to her lifelong friend, Elizabeth Haldane (1862–1937). Brought up at neighbouring Cloan House near Auchterarder, Elizabeth was herself an author, biographer, philosopher, suffragist, nursing administrator and social welfare worker and the sister of Richard Burdon Haldane, 1st Viscount Haldane, Secretary of State for War in 1905–12.

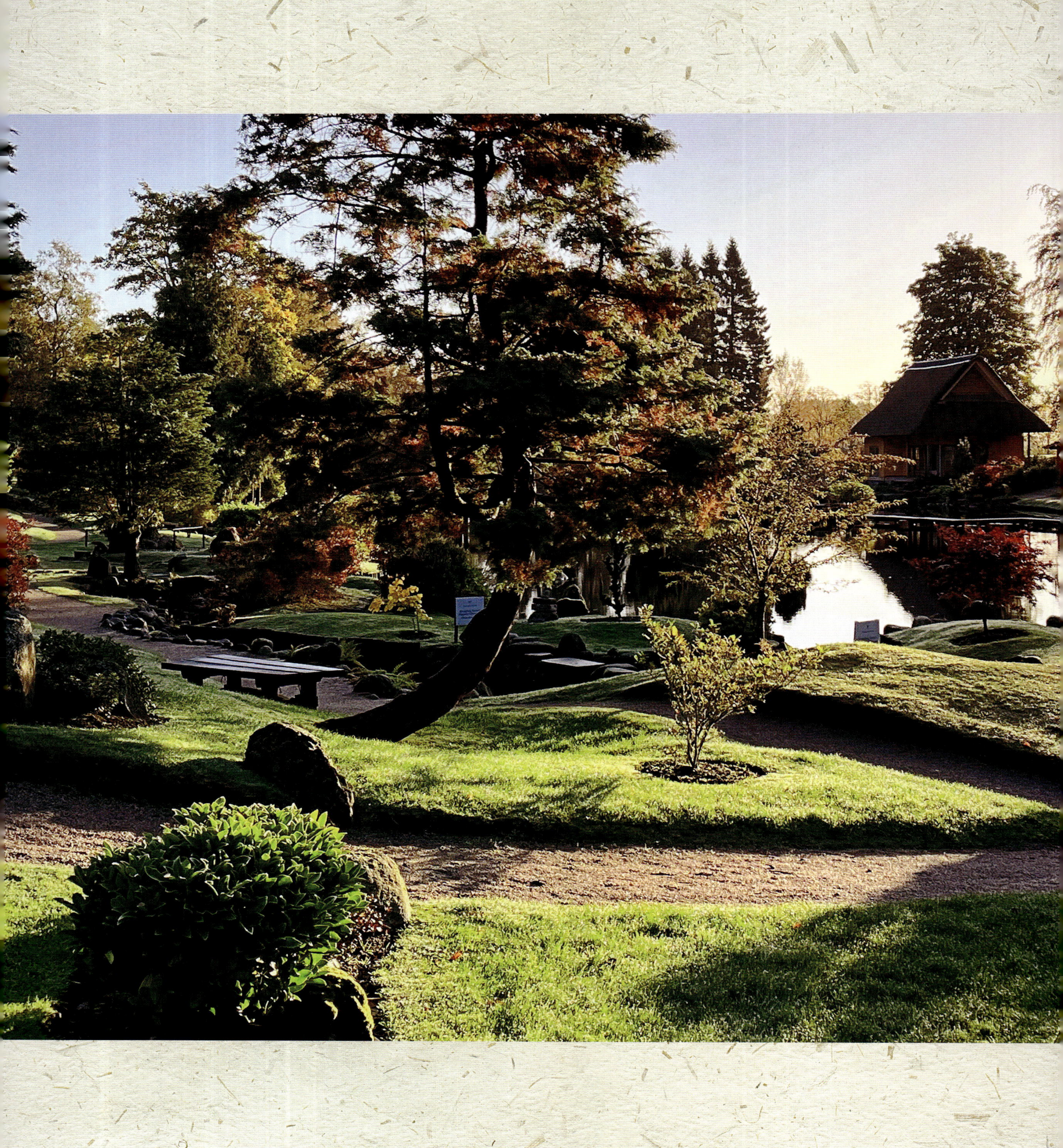

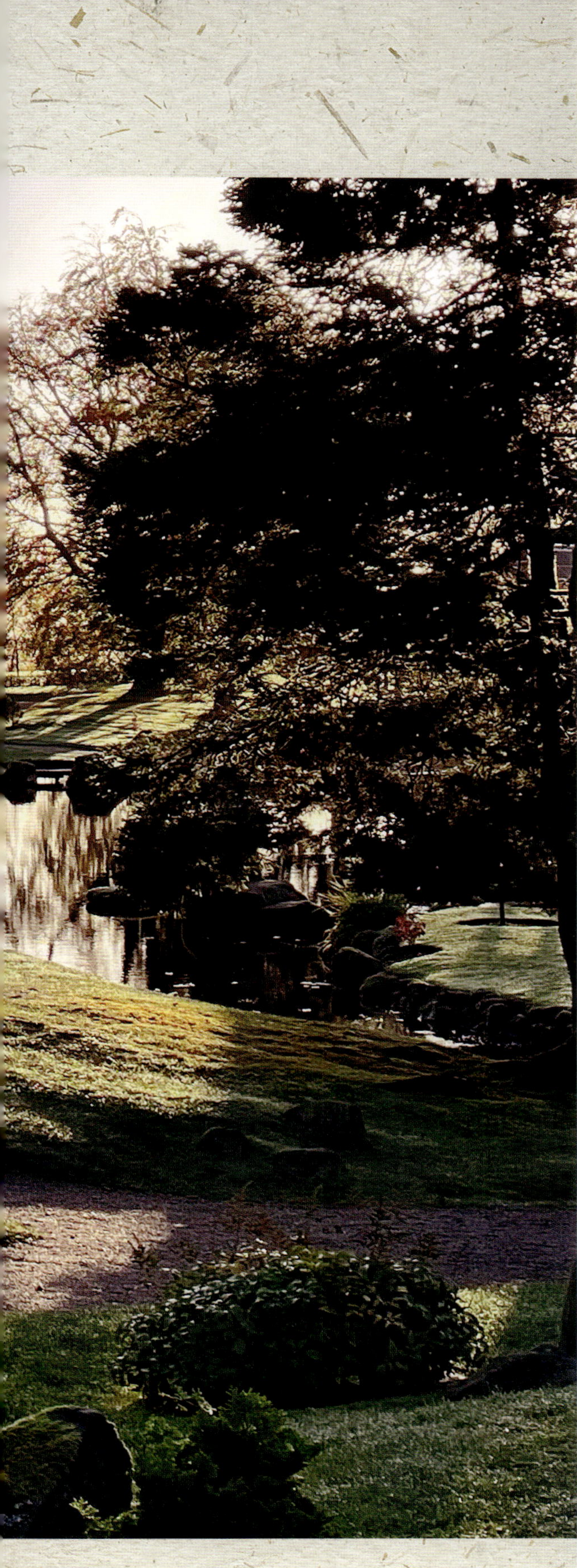

left Early morning sun over the garden.

4 · India and South-East Asia, 1904

ON JANUARY 8, 1904, armed with a parchment passport signed by the Lord Provost of Edinburgh, Ella set off from Marseille to Bombay on the SS *Mamora* with her maid Prudence Humphries. A redoubtable character of Warwickshire farming stock, Humphries would accompany her on several journeys, always travelling second class while Ella went first. Strong and wiry, she had 'a keen intelligent delight in travel,' and was 'fit to cope with any situation that might arise, however unexpected'. As a travelling companion, her company never palled, 'for she had a genius for recounting little incidents of the journey whether by land or sea'.[22]

Sailing via the Suez Canal and Aden, they arrived in Bombay in January 1904, where they stayed in a government bungalow and Ella attended various social functions, clearly charming everybody she met. Typical of a woman of her class, she had excellent contacts and the invitations flooded in – as they would wherever she travelled, so that she usually had somewhere comfortable to stay until she ventured off the beaten track.

By the end of January, she had travelled to Lucknow, Cawnpore and Agra, where she demoted the bearer she'd engaged in Bombay to bag carrier (she eventually

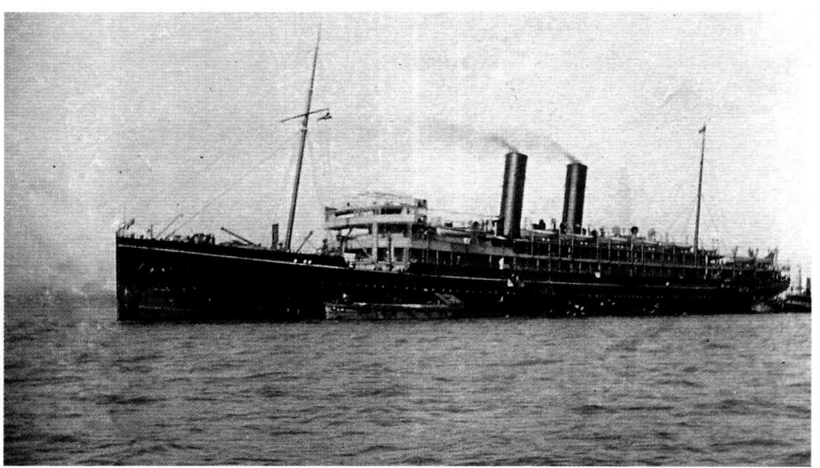

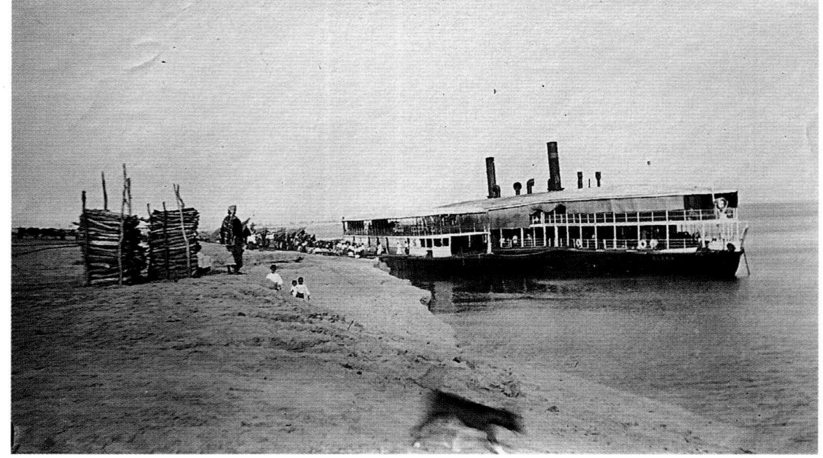

sacked him, complaining 'Sailoo is a perfect fool and knows nothing of travelling'.) Doing all the organisation herself, she continued via Delhi, Jaipur, Poona, Madras and Tanjore to Ceylon (Sri Lanka), where she had friends who ran a tea plantation. While in Nuwara Eliya, she was invited on a romantic rickshaw drive by an admirer, 'who got quite romantic and asked if I were in a hurry to be back, but romance has its limits in two rickshaws, even if the coolies are considerate enough to run side by side!'.[23]

By April, Ella was in Burma with a letter of introduction to a Mr Hay, secretary to the government, who made impeccable arrangements for her to travel on in great luxury. Arriving in Mandalay by boat on April 14, she made another admirer. M. D'Avera had been given the task of looking after her by his employer, the Irrawaddy Flotilla Company. 'Her manners, her talk and her unaffectedness, show that she had moved in High Circles, and one could see from the ease, the

gentleness and the bright spirit radiating from her handsome person, that she has been brought up among highly intellectual and refined associations,' he wrote to a colleague at the company's head office.[24]

Ella's next stop was Singapore, where she boarded a Chinese junk with a cargo of pigs – the only way she could get to Sarawak in Borneo, which she reached by May. She then crossed to Malacca and, having furiously ignored the discouragement of the district office, organised some transport, engaged 'a magnificent Sikh armed with a gun' and set off with Humphries up the Malay peninsula in a pony cart pulled by a bullock. They arrived at Penang, in the state of Kedah, in time for the Victoria Day Penang Derby. Here, Ella met the younger brother of the Sultan of Kedah.

They then travelled through Kedah, where five royal marriages were being arranged. Ella spent most of her visit in the company of the Rajah of Sarawak (Sir Charles Johnson-Brooke, who had succeeded his

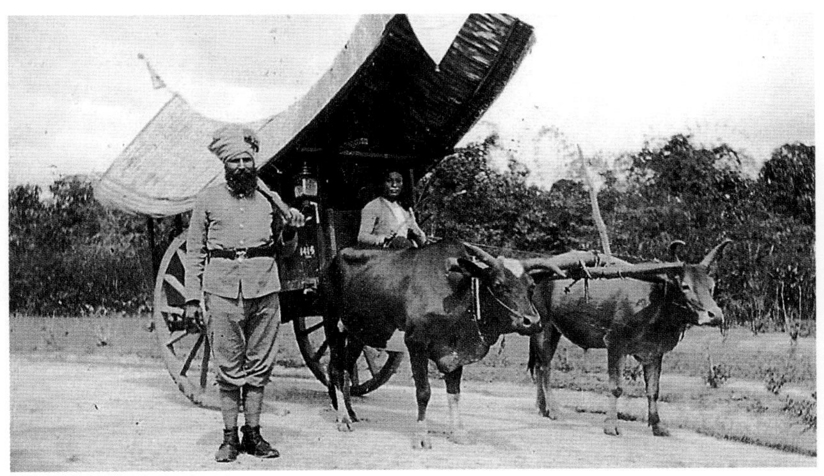

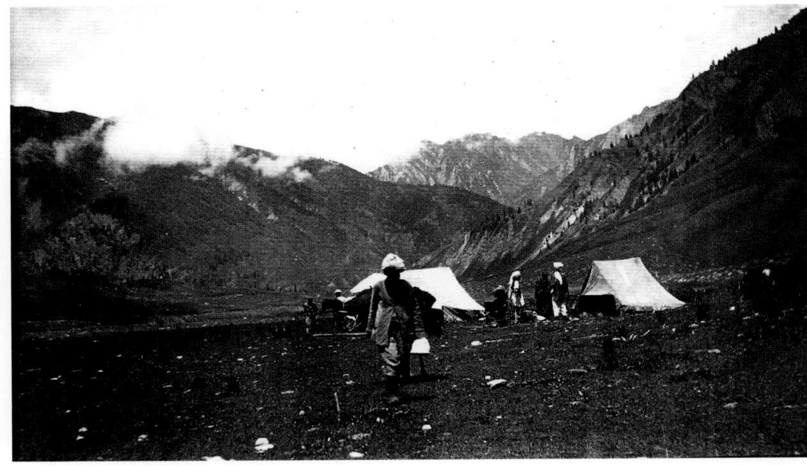

uncle James Brooke in 1868 as the second White Rajah of the territory in northwest Borneo, now a state of Malaysia). 'He has given me a fine signed photograph of himself with a request for mine, and is to take me to see his mother. As soon as I get my sarongs from him I shall make tracks for British territory as he has only filled up his complement of four wives allowed by Moslem law to the extent of one as yet.'[25]

After taking leave of Kedah, Ella and Humphries headed back to India, reaching Calcutta on June 6. 'Just arrived after a record passage from Penang for Badness ... it is a Jardine Matheson boat and is a disgrace to send passengers by it as it was never built for a passenger ship. Even H succumbed and said she was never so bad.'[26]

Later that month, they travelled up to the cool summer resort of Simla, where Ella made something of a splash. The season was in full swing and, on arrival, she received an invitation to a dinner being held in

honour of Lord Kitchener, then Commander-in-Chief, India. 'He began about the weather which after all one does not discuss in India, but I soon got him off that and we sat on a sofa and talked of other things,' Ella wrote in her letter-journal from the Grand Hotel in Simla on June 23.[27] There was talk amongst the local ladies; 'Humphries has been asked if I am travelling *incog.* and what is my real name? I have evidently been an enigma to the gossips'.[28]

In Gulmarg, further north, Ella left Humphries and continued alone to Srinagar, arriving on July 26. Her intention, in the face of official opposition, was to get to Tibet. She was not going to be daunted by the fact that the whole area was considered very much closed off and that no other foreigners had been there; indeed, this almost certainly increased the attraction of the expedition to her.

In Srinagar, Ella engaged an agent to source her camping gear and was scathing of his efforts: 'I don't

right A receipt from the Kashmir General Agency for items Ella stocked up on before setting off on her trek.

know what people do who trust to agents, as not even a teapot was there nor a jug of any kind and the debris in the kettle would have poisoned an oyster, while all the basins leaked'.[29] Other equipment she bought in preparation for the journey included a spine protector, a large *topee*, a filter and a short skirt about six inches from the ground. She also stocked up with a supply of drugs and joined the library, taking out eight books for 14 rupees on a three-month subscription.

Departure from Srinagar was by boat on July 29. 'Here I am fairly launched on my way to Leh, and so far nothing forgotten except a tablecloth'.[30] Further down the river, Ella disembarked and met up with a column of two ponies and 20 porters. The rest of her cavalcade comprised two permanent staff – her bearer Tikaram and a water carrier or sweeper – with porters engaged locally as and when needed. They were a troublesome lot, continually going on strike, but Ella proved their equal: 'as they had a month's allowance given on leaving Srinagar I declined to give more, and have said they may all go if they like … The Kashmiri is an awful beast to have anything to do with, and I only wish I had brought a revolver to brandish!'.[31] Eventually, she resorted to bribery: 'I announced three days ago that if there were no more complaints I would kill a sheep here, but if further demands, no sheep. This had an excellent effect and the sheep was killed this morning'.[32]

Ella and her troop climbed and crossed Zogi La at 11,000 ft. 'I walked quite twelve miles of the way, first because it was so perpendicular I could not stick on, and then [as] the down grade was so narrowly dangerous with a perpendicular descent to the Indus and no bush to catch one – worse than Castle Campbell.'[33] Protected from the strong sunlight by smoked glasses and *topee*, she walked miles each day, noting her breathlessness at such altitudes, but dealing with the food and the vagaries of the route with amazing cheer. 'One cannot be fastidious and one eats like a boa constrictor for the future. Today I began with three mutton chops as I knew I could only have light refreshment till eight o'clock dinner … Animal life is a drawback and one lives in an environment of fleas which Keatings has little effect on.'[34]

Srinagar Kashmir, 1st Nov 1904

Miss Christie

opposite left Unti se con consequi odignih ictusam, corene mentem net fugia por alit velectasin natur.

Bought of **KASHMIR GENERAL AGENCY.**

opposite right Unti se con consequi odignih ictusam, corene mentem net fugia por alit velectasin natur.

TERMS:—Bills payable on presentation. 2 per cent. per month will be charged on accounts over one month. Cheques on Indian Banks one per cent. commission. **CHEQUES SHOULD BE MADE PAYABLE TO PROPRIETOR.**

Date.		Rs.	As	P.
	To Last A/C Rendered	69	15	~
4-10-04	" Hire of 1 Tent + 1 Fall 25/9 to 4/10	7	~	~
	" 15 Iron Pegs missing	2	13	~
	" 1 Mallett	~	6	~
6-10-04	" 1 Tin Macaroni	1	6	~
19-10-04	" 1 Frying pan Broken	1	8	~
19-10-04	" Hire of Furniture 81 P.M. 27/9 to 10/10	6	~	~
	Total	89	~	~
	By 1 Tin Macaroni Bad 1/4/ 1 Pat-Butter -/8/	~	1	8
	Balance	87	4	
		88	8	~
	By Refunded on hire of tent	6	~	~
	Balance	82	8	

below left Where possible, Ella travelled by tonga in Kashmir, but much of the time she was walking on narrow platform roads made of planks of wood, suspended precariously above huge drops, or on stone tracks that were similarly rough and narrow.

below right On arriving at Khapalu, Ella chanced upon a festival that only happened every 35 years. One of the main events was a polo match.

On reaching Leh, Ella was unable to get permission to continue on to Chinese Tibet, being allowed only to go as far as Hemis monastery, 20 miles away. Reluctantly, she turned back to Srinagar, taking a different route. There followed a big argument with her porters, which eventually she won. 'I shook my parasol in their faces and said "price nay".'[35] Unknown to her, she was being observed by a pair of Englishmen out on a hunting expedition. She heard later that they had been about to come to her aid, when they saw her break her parasol over the head of one of the porters and decided that she had things under control.

The porters had a point, as the next step of the journey involved a hard march up to the top of Chorbat Pass, 16,700 ft high with a path that sometimes disappeared and air so thin that breathing was more like gasping. It was also incredibly cold. 'I hung myself with every available article of clothing, hugged a hot bottle, and drank cherry-brandy, but all of no use. I passed a sleepless night, and in the morning the ground was white with snow at least an inch and a half deep.'[36]

The descent involved 18 miles of tracks and platform roads – planks of wood supported on platforms held on cliff faces with a 400-ft drop to the river. On reaching Khapalu (now in northern Pakistan) at the bottom of the pass, Ella had another stoke of her customary good luck: she had arrived on the day of a festival that happens only once every 35 years. Furthermore, she encountered an official who took her to meet the Rajah and also met Jane Duncan, another intrepid traveller who had come specifically for the

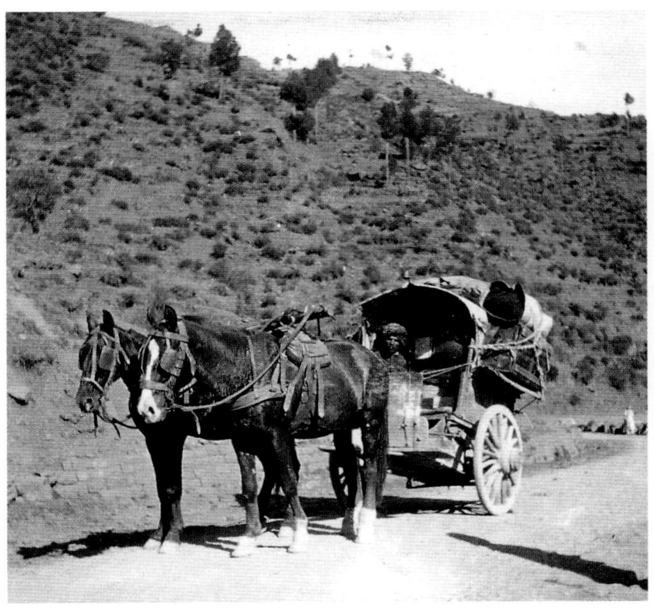

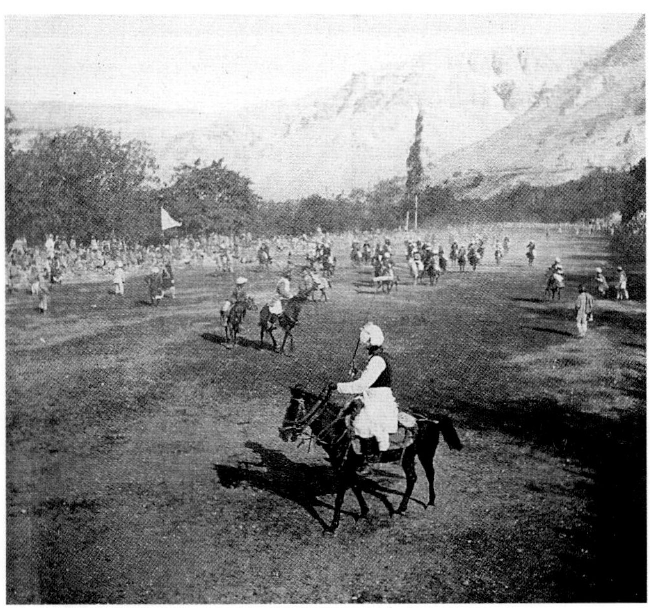

festivities. The only other European woman to have made it to these parts, Miss Duncan described the scene in her book *A Summer Ride Through Western Tibet* (1906): 'Looking around I saw to my astonishment, a lady walking into the bagh accompanied by the band and followed by an immense crowd. I was having tea and sent Aziz (her bearer) to give my salaam and ask if she would join me; she came at once and proved to be a countrywoman of my own on her way from Leh to Skadu. In coming down the hill she had fallen in with the procession in which she immediately made the leading figure and was immensely surprised, amused and delighted with her own dramatic entrance into Khapallu, and had no idea until I told her what it all meant. She too had camped at Goma Hanu and had had some trouble; the lumbardar refused to sell her any supplies and the coolies would not carry her baggage over the Chorbat La and she had to employ some women to take their place. It was bitterly cold and snowed all the time as she crossed the pass. She had met only two travellers on the road from Srinagar to Leh and from Leh to here, 400 miles in all'.[37]

Following a day of dances, riding displays and polo, Ella continued in biting cold along appalling roads back towards Srinagar, arriving at Skardu a few hours after Jane Duncan, having departed Khapalu a day earlier and endured a more challenging journey. 'She had gone down to the river intending to cross on the 'zak' (a skin raft) but there were a great many people from the tamasha going home by it who crowded on to it until they had to be driven back by main force,' recounted Duncan. 'After sitting watching it for two hours and seeing it sink twice Miss Christie preferred to do the next stage by the route marked as 'via Kurpak' on her map. Miss Christie found it difficult and dangerous, a mere goat track three of four inches wide in places across steep banks of soft, shifting sand, sloping to the river far below alternating with bands of clay in which notches had to be cut to give footholds. The coolies were extremely unwilling to go on after reaching the first village as the rest of the track was no better and wished to return to Khapallu but that would not have improved matters for it was equally dangerous to go back. She therefore pushed on and arrived at a miserable camping ground beside a miserable village nearly opposite to Kiris so worn out that she had to rest there a day before proceeding further. No European woman and very few Sahibs had ever used this route and the villagers came in crowds to look at her.'[38]

When Ella eventually reached Srinagar, by way of the Deosai plateau, the Gilgit road into Kashmir, the Tragbal Pass and a boat from Bandipur, she was reunited with her maid and they set up residence on a houseboat. With the help of some furniture and some blinds sewn from white muslin turban material purchased in the bazaar, they made the ss *Heliotrope* quite comfortable and Humphries attempted to make her mistress presentable in anticipation of a state banquet to be given by the Rajah on November 22, in honour of the Viceroy. 'You seem to be gnawing a crust by the roadside one day and in government house the next,' Alice observed to Ella in a letter. The evening was like nothing Ella had experienced before.

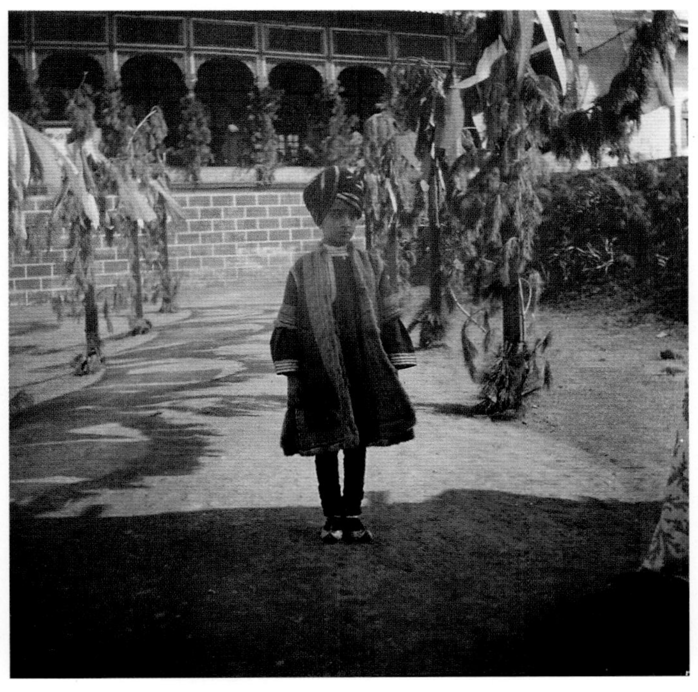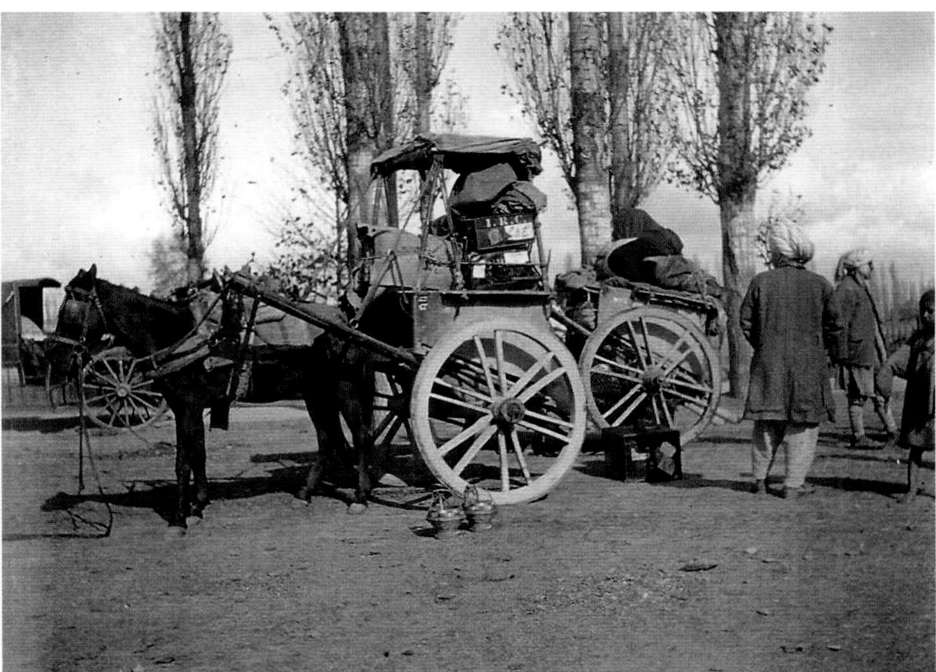

It took place in a huge palace lit by flaring torches, to which 200 guests were invited, each being introduced to the Maharajah, who sat on one of a pair of gold and silver thrones surrounded by throngs of retainers. 'He is about 4 feet nothing, half swallowed up in an enormous white turban, roving eyes and usually under the influence of opium after 2pm.'[39] The guests were served by attendants, who 'made irregular dives at people with the dishes, so one had to take whatever was offered for fear of getting nothing'. There were 'excellent' speeches and fireworks and, the following day, 'several have been ill today from the effects of the feast'.[40]

After this excitement, it was back to the challenge of getting to the next destination, Islamabad. They arrived there, via Peshawar, on December 1 and Ella 'got some washing done as the dhobi has an iron. You cannot think how much that means'.[41] As it was out of bounds to Europeans, she merely looked up at the Swat valley, but she did then visit the Khyber Pass, before continuing on to Rawalpindi, Agra, Jodhpur,

Amritsar and back to Bombay, where they caught the steamer home in January 1905. Ella had been away for a year.

Despite all the problems with the porters and the difficulties of getting accurate travel information from anybody, official or unofficial, Ella very much enjoyed her time in India and her interest in the culture and love of the scenery would remain undimmed. As one of her correspondents, Sir Frank Drummond, wrote to her, 'It is a pleasure to find anyone who comes out here so appreciative as yourself. Is it not wonderful how little people here know of the beauties, both architectural and otherwise – and I can assure you they know as little of the inhabitants … in you I have found a rarity. If more of your appreciation was applied to India and its administration one would hear less of our unpopularity and perhaps even gain the gratitude of its people'.[42]

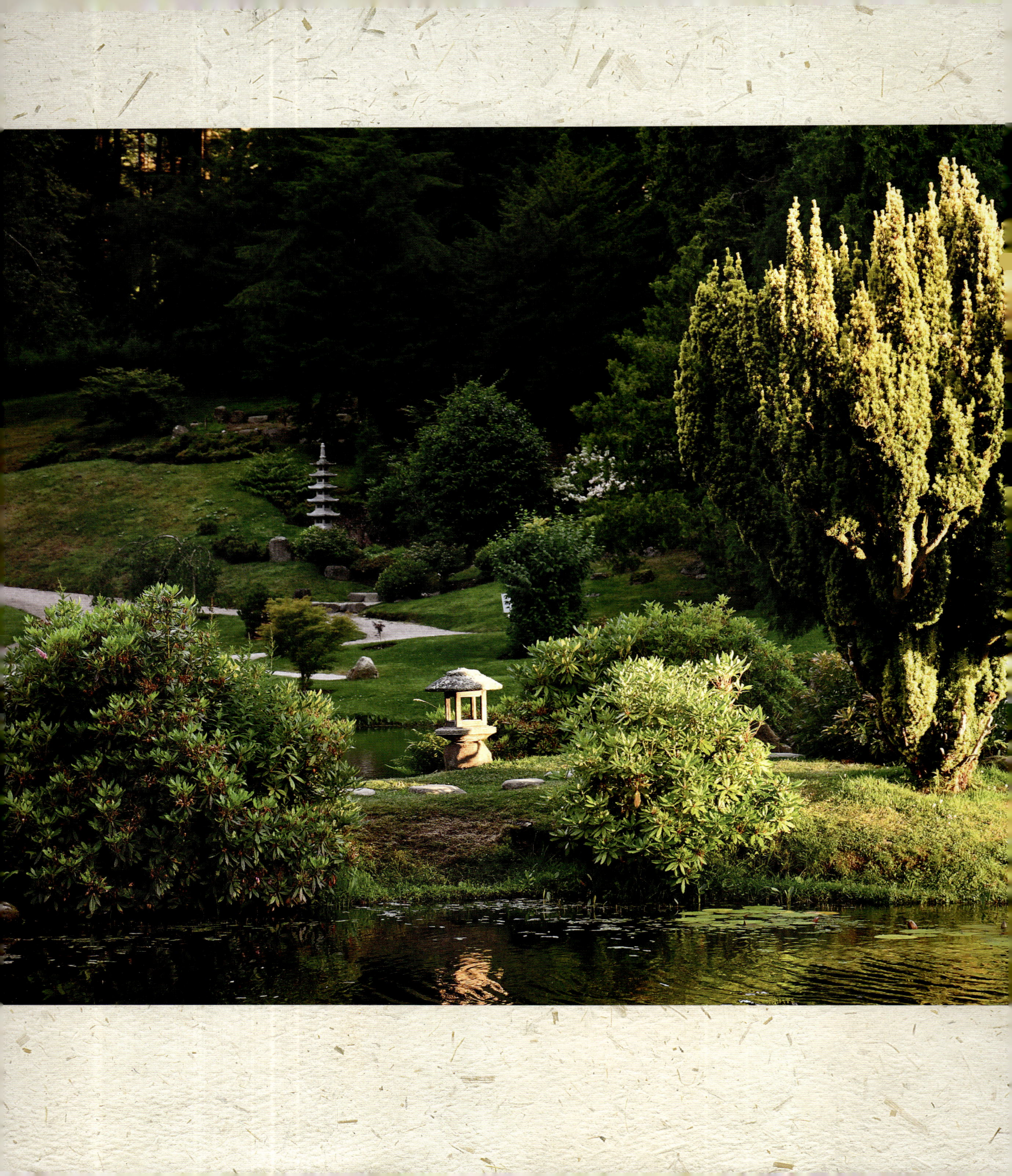

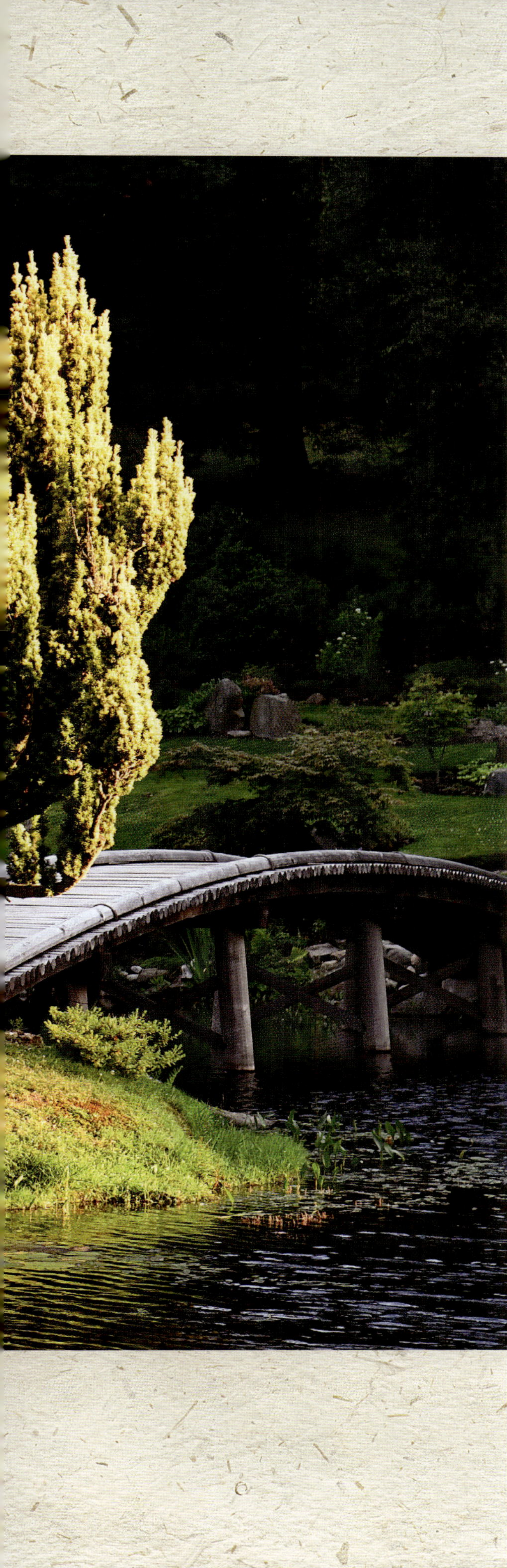

left The Korean Pine on the *nakanoshima* was grown from seeds that Ella brought back from her 1907 trip to the Far East.

5 · The Far East, 1907

ONE MORNING IN 1906, while waiting on the platform at Dollar Station in her homespun tweeds and sensible hat, Ella bumped into a friend, who asked her if she was going to Edinburgh for the day. 'No, to Japan for a year,' she replied.[43]

The trip had had to be delayed, but the Russo-Japanese war (1904–05) had recently ended and so, at last, Ella was on her way, armed with letters of introduction, maps and documents and accompanied by the redoubtable Humphries. They sailed via Colombo, Penang and Singapore and arrived in Hong Kong in February 1907. Having made it her business to get to know everybody on board, Ella was immediately swept up in a whirl of lunches, dinners, the races (her visit coincided with the Hong Kong Derby) and trips to Macao – 'a very pretty peninsula so clean and bright looking, the houses (European) painted cream or pale lavender' – and Canton. 'I never expect to see anything stranger than Canton,' she wrote to Alice, describing with a mix of horror and amazement the myriad handicrafts, stomach-churning food markets and general stench and filth of the labyrinth of narrow streets, not to mention the noise 'partly from coolies with heavy loads slung on bamboos, keeping

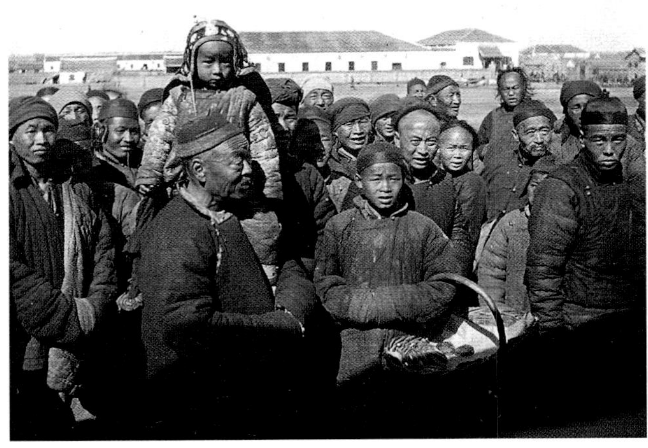

top An inquisitive crowd at a station between Hankow and Peking.

centre Ella was fascinated by what she called the native city, as opposed to the foreigners' legations. Here, she photographs a Peking street thronging with life.

bottom A roadside restaurant in Peking.

up a simultaneous grunt and trot, shouting to clear the way'.[44] In Hong Kong, she was shocked at the 'deformed feet' of the women, but admired their 'rational' dress of 'wide trousers, usually black, and sack coats' and thought 'the tiny girls with a *chignon* behind the left ear are deliciously quaint'.[45]

Early in March, they sailed up the coast on the ss *Prinz Ludwig* to the mouth of the muddy Yangtze River, where they embarked on a launch to steam up the tributary river Huangpu to Shanghai. There, Ella's Scottish thrift kicked in: 'None of them [her fellow travellers] ever bargain nor ask the price of rooms, which of course is idiotic but truly English. Before ever I stepped upstairs in this hotel, I enquired the price of the rooms and was told "ten dollars" … I offered four for myself and three-fifty for Humphries. There is no difference for maids. The money is maddening, a different kind at every place in China'.[46]

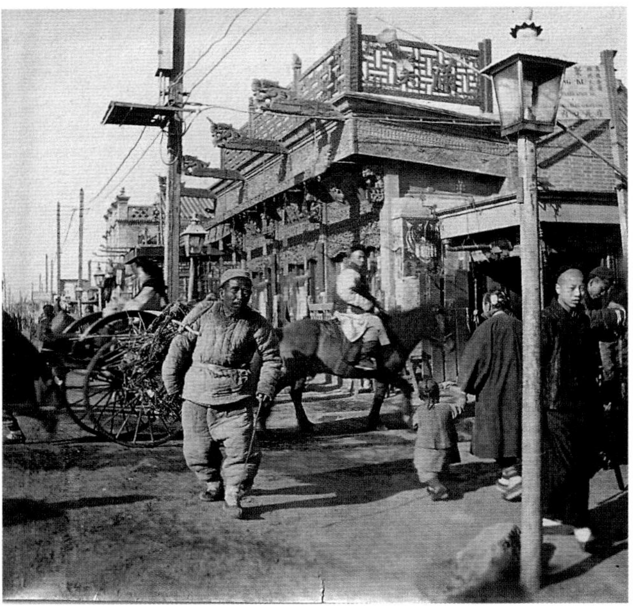

Having again networked extensively on the boat, she was entertained by her new friends and taken to the races. She was very impressed with how organised the Japanese were in such things as the arrangement of the national concessions in Shanghai. She described to Alice the great promenade along the river bank called the Bund, the bustle of the traffic, bicycles dashing like lightening and 'coolies' in pairs shouting for a free passage while carrying their heavy loads slung from a bamboo pole. A day spent in the overcrowded walled city left her of the opinion that a European would be unlikely to survive for long in conditions that were as fetid as those she'd encountered in Canton. As ever, she shopped and bargained for interesting things to

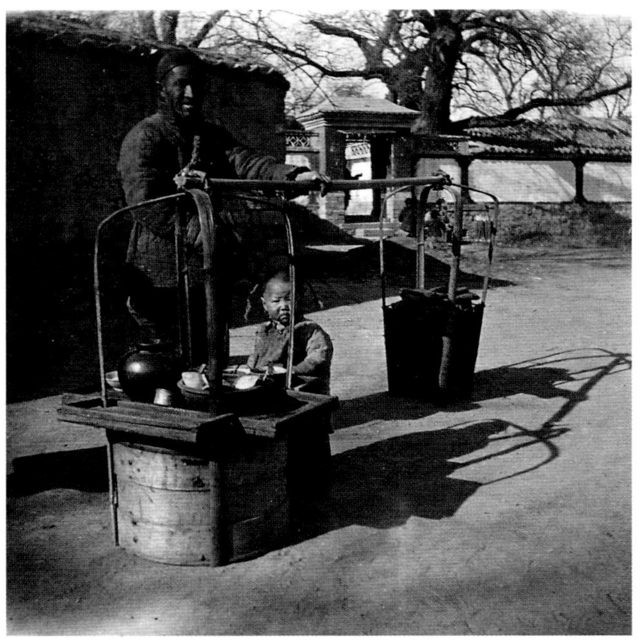

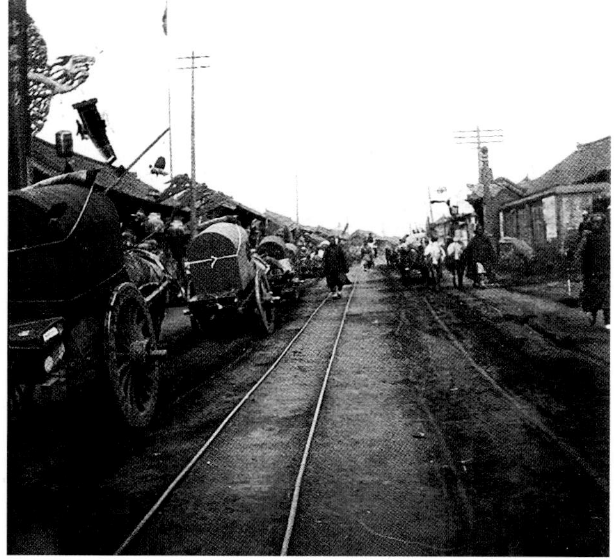

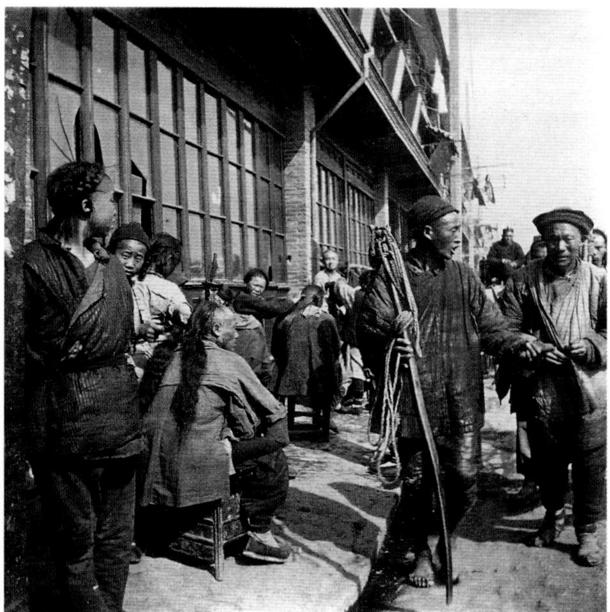

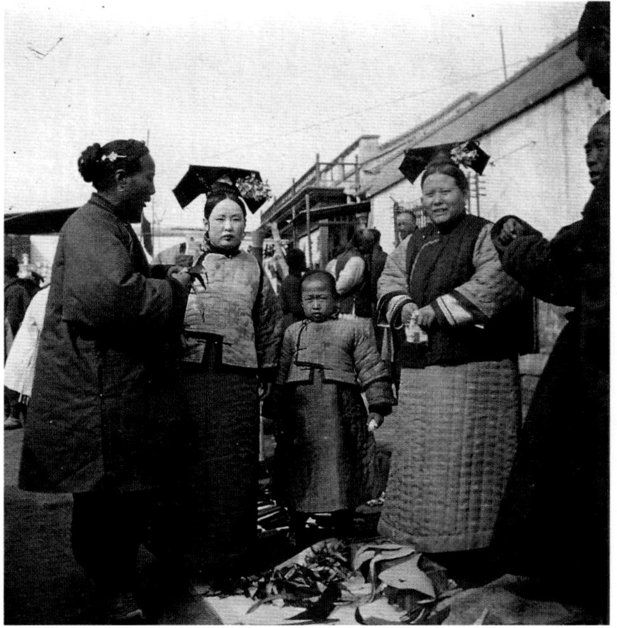

top Mukden in Manchuria, with carts transporting luggage belonging to Ella and Miss Humphries.

centre Roadside hairdressers in Peking, where the Chinese had their queues (long plaits) rebraided.

bottom Manchurian ladies sporting ornate hair arrangements in Peking.

send back to Cowden. She visited the jade street and the silk street, marvelling at the ivory carving and metal work displayed in open-fronted shops, the abundant food shops and restaurants, the original of the willow-pattern tea-house, the garden of some wealthy mandarin and the Confucian temples.

They then took a four-day journey in a comfortable river-boat up the Yangtze – 17 miles wide at this point – through flat country covered with rice fields and straw-thatched huts to Hankow. From there, it was 38 hours by train to Peking (now Beijing), where Ella presented her introductions to various contacts. 'I left my letter at the British Legation and this evening the British Minister himself arrived to call. Which I call civil. He is Sir John Jordan and first knew Pekin [*sic*] 30 years ago. He asks me to lunch at 1 o'clock tomorrow, and he will show me all around the place.'[47] The 'strings of Bactrian camels, with Mongol attendants, Manchu women in gorgeous head-dresses tottering about, [and] carved shopfronts in such wonderful designs' dazzled Ella and Peking also further satisfied her magpie tendencies. 'I started on the curio shops this afternoon and have got a pair of Pekin boots, a large enamel plate, four snuff bottles, and an old blue vase. I can see there are good things to be had.'[48] However, she was shocked by the damage done by Anglo-French forces in 1860, during the Second Opium War – the smashing of the Temple of Heaven's blue and green tiles and the destruction of the Summer Palace – and also by the damage to the Golden Temple in the Forbidden City, which the Japanese had attacked several decades later.

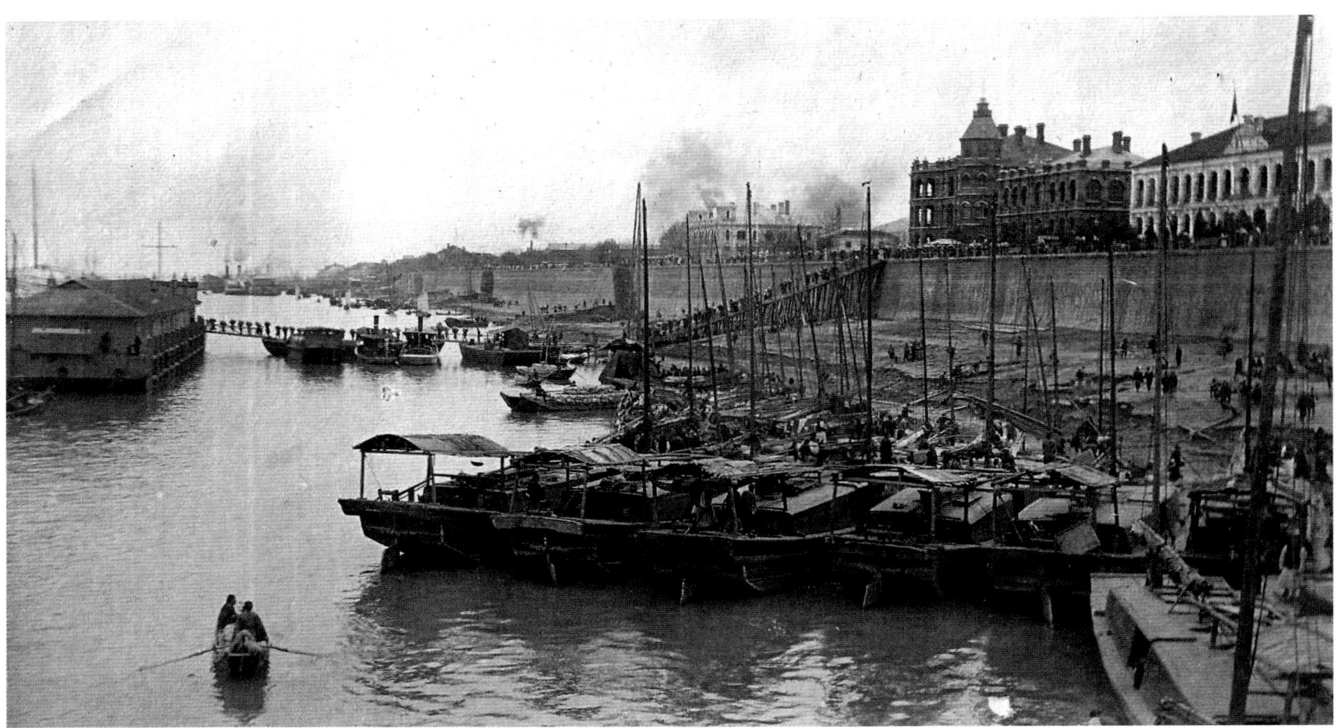

On March 26, Ella wrote from Tientsin (now Tianjin) to Alice: 'My Dear A ... This day has been a record snow storm from early morn but I had to go out to make arrangements for my journey north ... So, I have passed customs, changed money at the Japanese bank, got a free pass for all Manchuria, had tea with the Jap Consul, and last but not least been presented with a box of jujbe tree fruit, commonly called Tientsin dates. I must say the Japanese people are civil beyond words.'[49]

The mode of travel now became increasingly basic as the two women continued by train to Mukden in Manchuria, until recently a war zone. The station master directed them to the best Japanese-run hotel – Japan had now replaced Russia as the dominant influence in the region. 'We had shelter for the night certainly but no more. The beds a species of shelves or boards, were filthy so we lay on the top, and lived in the tea basket for supper and breakfast. I am travelling at present in addition with cheese and walnuts.'[50]

There were a few exceptions to this rudimentary existence, for Ella was invited to a candle-lit dinner by the Japanese Consul based in Mukden, a man named Hagiwara. 'We began with pine-seeds and a dish of salt strips of something in vinegar, then pounded chicken and little bowls filled with bits of

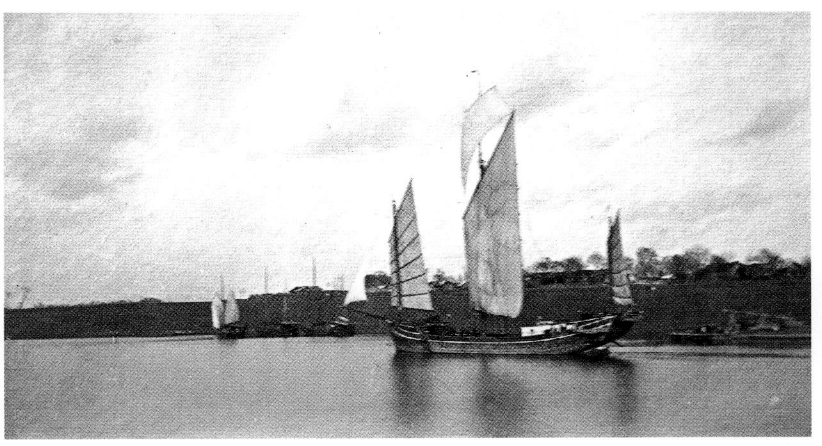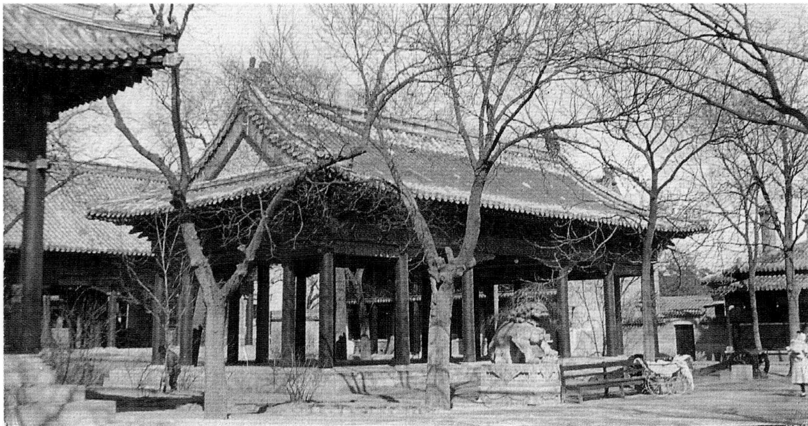

fish, potatoes, vermicelli, and greenery, then roe with soy, bean custard, and what looked like slugs. This last I did not touch. I got quite agile with my chopsticks towards the end. Claret and weak tea was served, with port afterwards.'[51] Humphries, meanwhile, was cheered to learn that they were only 16 days by train from London.

Their next destination was battle-scarred Port Arthur (Lushun, China), where they were greeted at the station by the local chief of staff's ADC, the Mukden Consulate having notified him by telegram of their imminent arrival. 'Is the lady then *not a British general*?' inquired the ADC. Ella: 'Alas, I am not.' ADC: 'Perhaps then the lady is *widow* of a British general?' Ella: 'Not that either – but there is no saying *what I may become.*' ADC to Humphries: 'Is the lady's name *really* "Miss Christie"?' Humphries: 'That is the name she wishes to be known by.'[52]

Put up in the best suite of the only European Hotel, which was next to the pigsty, the ladies were presented with garlands and whisked off on a tour of the battle-fields. But then a disaster occurred. Humphries was in a cart in Port Arthur when the horses bolted and she fell out, landing on her head. She was taken to a Red Cross hospital run by the Japanese, where 'the doctor, a little Jap did not know a word of English and a very limited amount of German, however with that one

managed. He was filled with importance and always kept referring to excoriations on her forehead which meant nothing to speak of … I stayed all night and most of today and this morning she was quite herself but knows nothing of the cause of her illness, so I have not told her. She still feels her head and is strained and bruised.'[53] The nearest possibility of proper medical attention was the American hospital in Seoul (now in South Korea). Ella undertook the nightmarish journey with Humphries, travelling for seven hours in a covered train truck to the port of Dalien and then boarding a boat, where a kind Russian engineer, whom Ella had befriended in Port Arthur, appeared with a bag of food to see them through the three-day voyage.

On reaching Seoul, Ella recovered from what she would later say was the worst journey she ever experienced by setting off to explore with some recently made friends, while Humphries was left under medical care. 'We are in a comfortable hotel and I got an American doctor this evening who assures me there is no injury to H., and chiefly nerves, so by a week he hopes she will be alright'.[54] By April 17, they had crossed from Pusan (Korea) to Miyajima near Hiroshima, where she was clearly still feeling somewhat frustrated: 'H's accident has upset my plans owing to the delay and I cannot bustle her about as she is still suffering from the shock to her nerves'.[55]

Despite Humphries's frail state, however, it was not long before they were continuing their tour of Japan. They proceeded to Kobe, where Ella met up with Ella Du Cane, whom she'd met on board the *Princess Alice* sailing from London to Hong Kong. Miss Du Cane was a successful painter, who had come to Japan to do a series of watercolours of gardens. She described Kobe as 'really a sort of Greenock' and advised Ella that there was no need to waste time there; she should get on to see the riches of Kyoto as soon as possible. She then wired her sister Florence, who was in Japan writing *The Flowers and Gardens of Japan* (the book that they would jointly publish as author and illustrator in 1908), and asked her to secure a room for the travellers in Kyoto. The sisters were an important part of the jigsaw; they acted effectively as Ella's hosts in Kyoto, guiding her around the gardens and introducing her to all the people they knew there.

Ella's arrival at Kyoto coincided with *hanami*, when the Japanese go out to view the cherry blossom in all its glory. 'The blossom is a wonderful sight, cherry both double and single, and soon the azaleas will be out,' she wrote to Alice from the Yaami Hotel in Kyoto on April 24. She was also very taken with a demonstration of *ikebana*, Japanese flower arranging. 'That really is artistic and so interesting to see a man in front of a bronze vase or dish pruning and twisting and turning and snipping perhaps only a pine branch, till it is arranged to perfection.'[56]

There was still so much more of Japan to see. Ella engaged a guide and took him up to Tokyo, but the 'gentleman boy', as Kanaya at the Nikko Hotel described him – 'he has appeared in 3 different brocade waistcoats, the June 4 is not in it with him,' Ella joked – was 'useless' and she sent him back to Kyoto. Instead, she engaged 'a nice rickshaw boy,

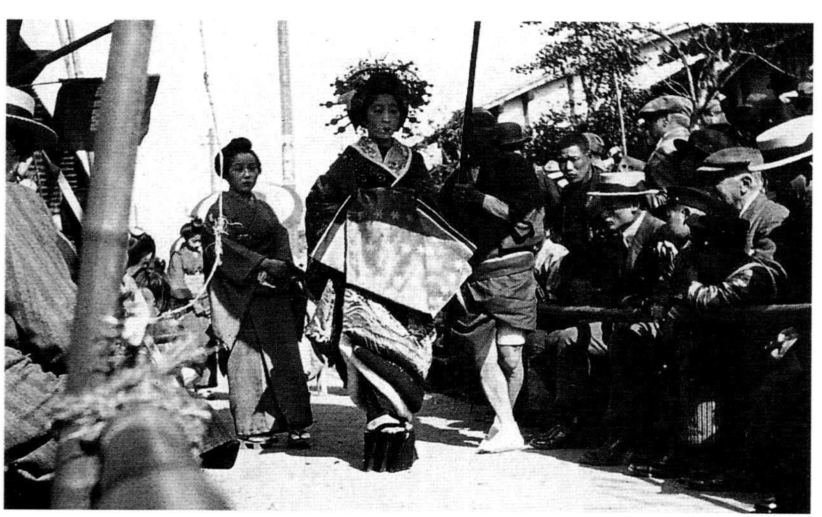

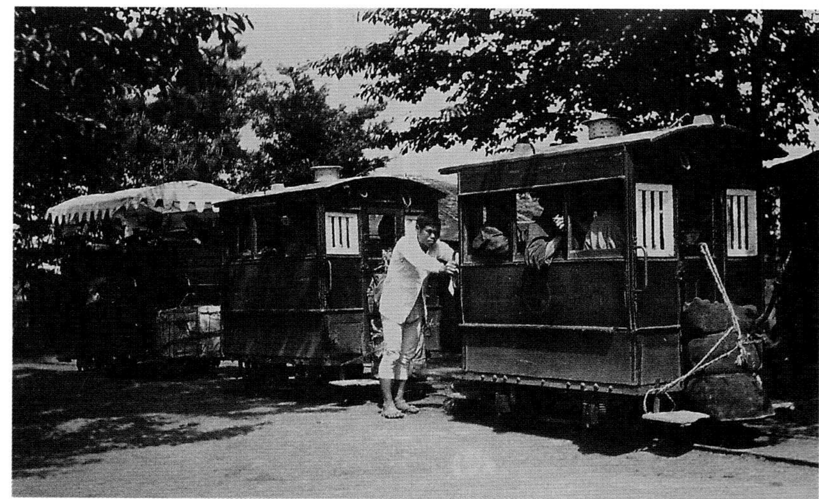

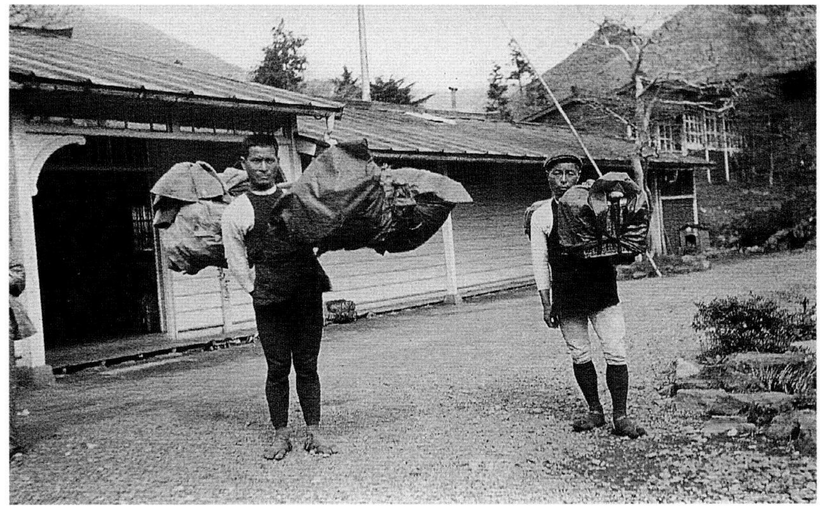 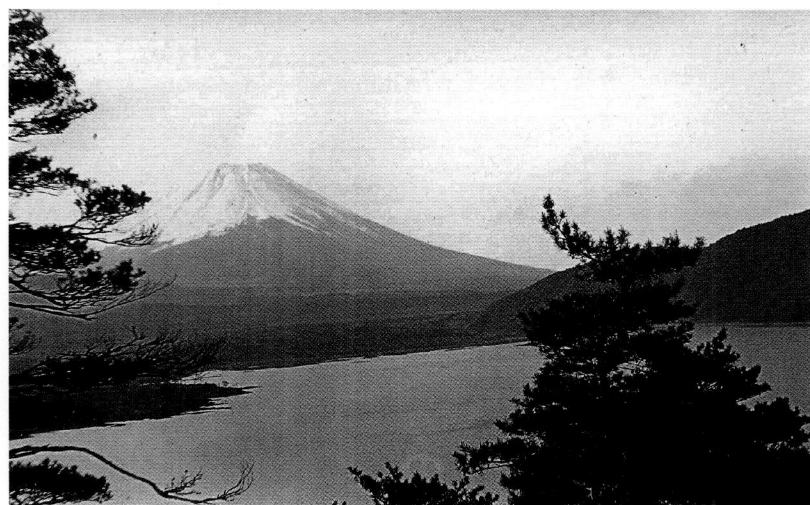

Boon by name, who knows a little English and is most willing'.[57]

The two Ellas travelled together around the Tokyo hinterland, staying in local inns. 'Miss Du Cane is a capital traveller and most handy, one of the few women I have found one can travel with as she is capable. I left Humphries in Tokio [*sic*]. She has no love for Japan. Not the excitement and delight of India, she finds'.[58] Ella found certain things – the hot spring baths, for example – a bit odd, and wrote dismissively that 'you can't believe a word a Jap tells you as to distance, a missionary said to me, it is just their way, they want to please. "Well," I said "it would please me more to hear the truth".'[59] Some of the food was a challenge, but her companion travelled with a hamper so that they could supplement the less edible local food they were offered with their own supplies. In spite of all this, she greatly enjoyed her time in Japan.

In early June, after a 'wretched sail' to Vladivostok – 'a bleak place with a fine harbour, reminding one of Port Arthur' – Ella and Humphries took the Trans-Siberian Railway to Moscow, travelling through luxuriant primeval forests, with swathes of wild flowers along the tracks, and the 'bleak wind-swept plain' of Siberia, through which emigrants were being herded

east in train carriages worse than cattle trucks to escape the famine districts around Kiev.[60]

Both the railway journey and the Russians left a bad taste. 'The carriages are not bad but the arrangements are bad in the extreme and the food vile … From what we have experienced so far the Russians seem half savages in their ways,' Ella wrote from the train on June 15, adding 'the Duma has been rejected and I am advised to leave Moscow as soon as possible as no one knows what will now happen'.[61] After a chaotic three days, she managed to arrange passage on a train to Paris. By now looking decidedly ragged, she reached the French capital with a sigh of relief and immediately set out to buy some new clothes.

This was the last trip that the two women made *a deux* and Ella must later have missed her maid's company. In many ways, Humphries was an ideal travelling companion, interested in what she saw, intelligent, free of speech, full of amusing turns of phrase and tough as her mistress. Ella's next two journeys through Central Asia were to be made solo.

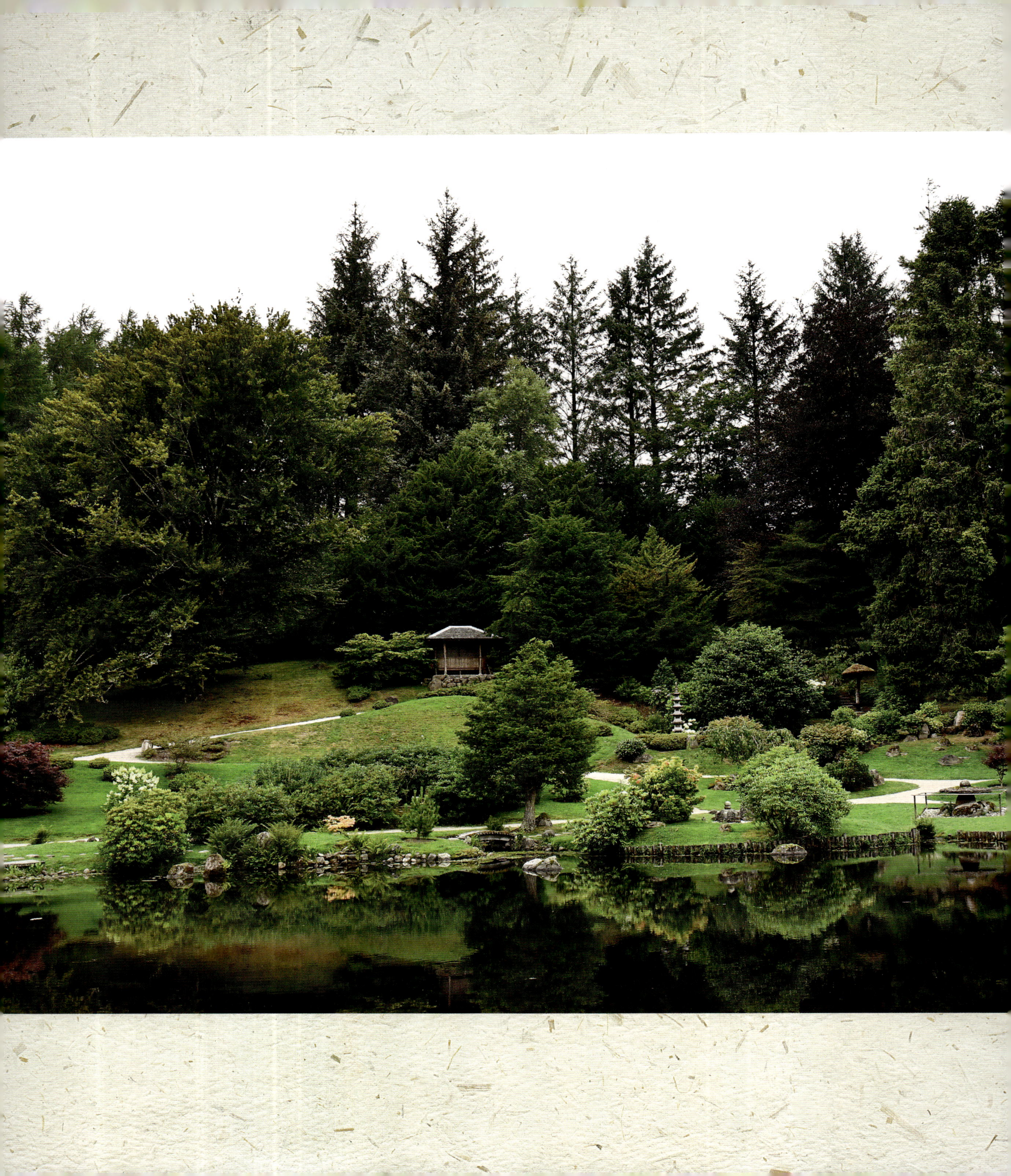

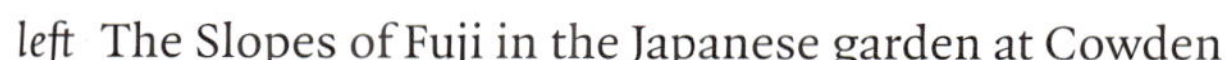

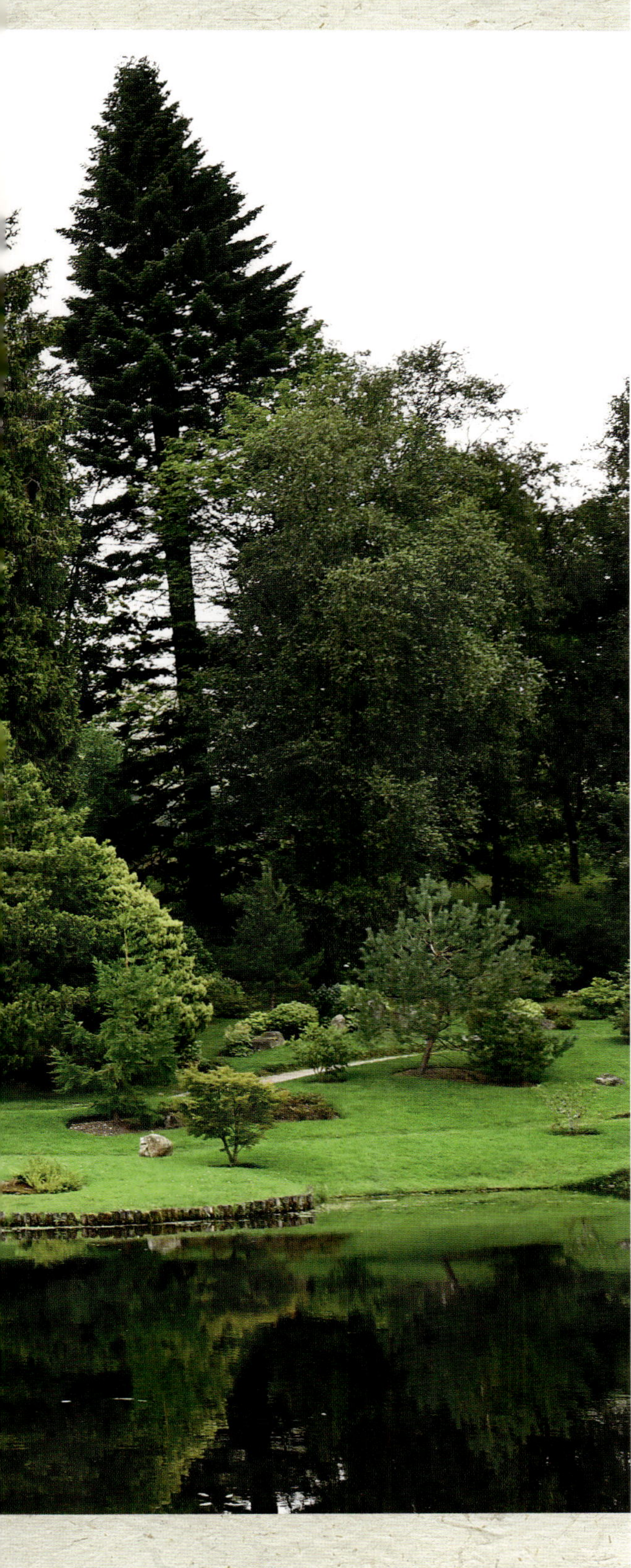

6 · Central Asia, 1910 and 1912

FEW FOREIGNERS UNDERTOOK the challenges of visiting Central Asia when Ella travelled there in the early 1900s. A letter from the Foreign Office made clear the dangers: 'The sanitary condition of Russian Central Asia and Bokhara are not satisfactory at this moment. There is some plague and much small-pox (both the ordinary and the black variety), scarlet and typhus fevers and diptheria. There is also a dearth of vaccine and competent vaccinators'.[62] The infrastructure was virtually non-existent and there was also the risk of being accused of spying. The long-suffering Colonel Haldane, Elizabeth's cousin, who was in military intelligence at the War Office (he would later become General Sir Aylmer Haldane), was pressed into action again and, through persistence, Ella managed to get permissions from the Russians, a *laissez passer* from the foreign secretary Sir Edward Grey and an impressive document bordered with blue dragons from the Chinese officials. These she packed along with bedding, a camp bed, some insect repellent and a bag of oatmeal.

Ella was accompanied on the first part of the journey by Humphries and 'M'.[63] The ladies set off from Charing Cross Station on February 25, 1910, got to

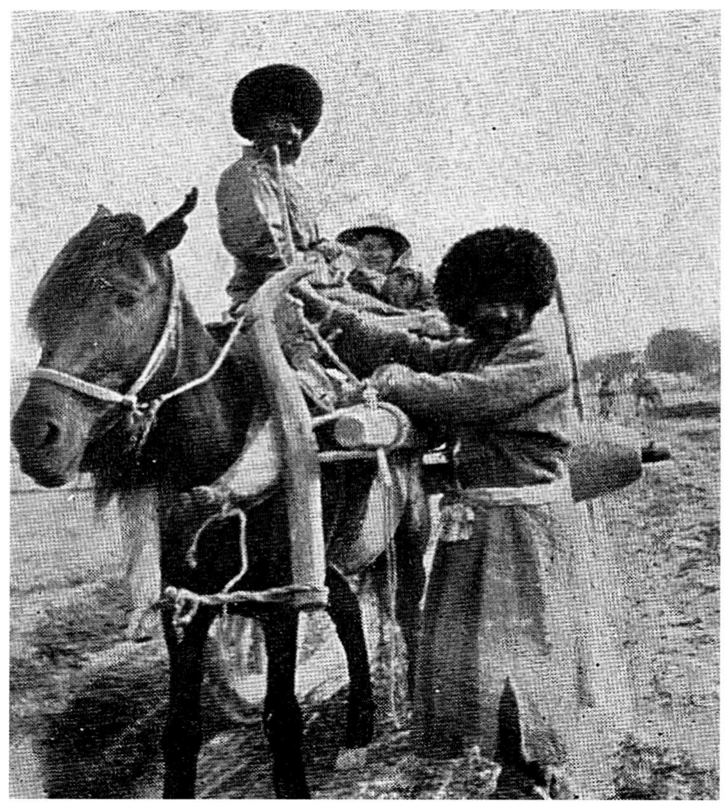

Budapest 36 hours later and took the Orient Express
to Sofia, arriving in time for Bulgaria's Independ-
ence Day celebrations. 'Our visit could not have been
more successfully planned – Ella has good fortune
wherever she goes. It is always the right time and the
right place.'[64] Having called in at the British Legation, they were duly invited for lunch and to see the
parade, where, tourists being unknown here, they
were mistaken for ladies from the legation and given
special seats.

In Constantinople (Istanbul), letters of introduction ensured that they continued to be entertained.
Everywhere they went, Ella was introduced as the
lady who was going to Bokhara 'and each honoured
guest pressed forwards to get one word – one shake
of the hand, some treasured look from the celebrated
traveller'.[65] They visited Scutari, took boats along the
Bosphorus and explored Istanbul, visiting mosques,
markets and hospitals.

Advised by Sir Donald Mackenzie Wallace, writer,
editor and foreign correspondent at the *Times*, that
she should boil every drop of water she drank and not
even risk washing her hands unless the water had been
boiled, Ella added a samovar to her packing. She also
decided to get herself vaccinated against smallpox and
had the injection on March 23. This left her feeling
decidedly ill for the rest of her stay in Istanbul and she
was still in bed on the 28th, when she had to get up.
That afternoon, at 4pm, Ella parted company with 'M'
and Humphries at the place from where her boat was
due to sail the following day. 'As it was getting late,'
'M' wrote in her journal, 'it seemed best not to wait

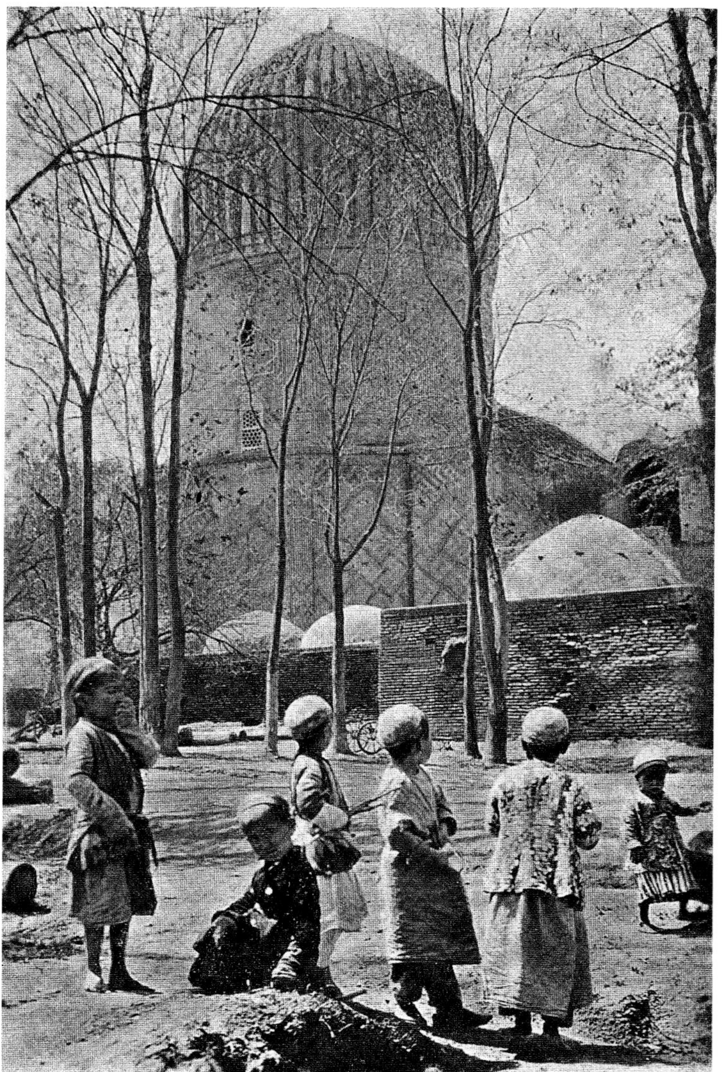

as Humphries and I were heading west that evening so with much regret I left Ella on ss *Louie* and our five happy weeks were ended. I could see Ella at a great height, alone and it seemed very sad and very flat as I turned to wave at her and call out "Mashallah".'[66]

The ship on which Ella had taken a passage belonged to a French coal company. After a three-day journey, it arrived in Batoum (in modern Georgia), where Ella stayed in the Hotel Imperial and dined with the British Consul. Through his services, she secured a guide called Abuloff, whom she described to Alice as 'dark and dirty and goes about with a white flannel tied round his face … an Adelboden hood and a yachting cap, but I think he is decent'.[67] They took a train to Tiflis (Tiblisi), where they called on the Russian Consul, and then continued to Yerevan and Kirovan. Thanks to her letters of introduction, Ella was royally entertained by the governors and she handed out postcards of Scotland and pebble brooches, which were well received. Having made a side trip by horse and cart up into the mountains, she decided to revert to trains, which seemed a safer option. By the time she reached Baku, she had resolved that 'Abuloff is too great a fool for words, knows nothing of trains nor of anything else. If I can get an improvement on him here I shall change'.[68] And so another guide bit the dust and, with the help of the British Consul, she found the Jewish converted Russian Abram Ovitch, who had spent 20 years in Australia. Abuloff was furious at being sacked and 'I fully expected him to draw his revolver, so carefully left the door open while he strapped up his luggage, and with any suspicious movement of his hand towards his pocket was into the passage. He followed me to the boat, pouring curses on me, and sat beside me waiting for my Jew till I told him he need not incommode himself further as I felt the night air was bad for his neuralgia! An Armenian will stab you at once for any fancied injury'.[69]

Ella then travelled by way of Askabad and Merv along the Silk Road as far as Andhizan, all the time taking photographs with her Kodak camera and being entertained by an assortment of officials.

At Bokhara, she wrote to Alice: 'From the moment of arrival I feel I have been witnessing some wonderful pageant. It is the extraordinary colouring of the clothing that gives the charm – and not a European to break it'.[70] Ella had read all she could about this fabled capital of Tamerlane and describes the walled town in detail in her book, *Through Khiva to Golden Samarkand*. As she did on arriving at any new place, she went shopping, nosing around the arcaded bazaars, buying some astrakhan skins, old brass and other curios, and visiting the caracul market, where the lambs were sold and then slaughtered, the saddlery, dyers' and shoe bazaar and the famous silk quarter. She watched scenes of medieval splendour outside the 12th-century citadel (which, as a woman, she could not enter) and the Emir, mounted on a bejewelled horse making his way to a service in the mosque, which she attended and photographed – a sight rarely seen by a European woman.

By April 29, they were in Samarkand, where Ella noted 'They are great sweet eaters and all sorts of confections are prepared from mutton fat'. She marvelled

 On Ella's second trip to Central Asia, she took the train from St Petersburg, travelling second class because first class was occupied by the Russian Minister for War, General Sukhomlinov. Every stop featured some sort of formal welcome for him.

 The train journey from St Petersburg to Tashkent took five days.

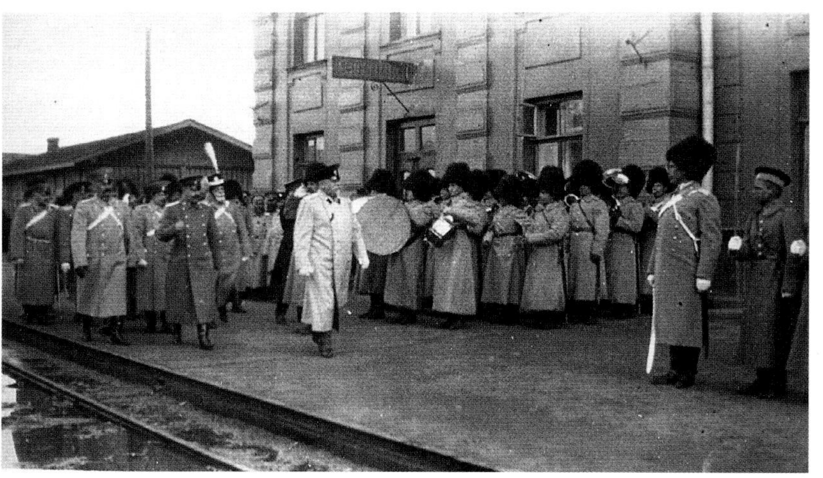

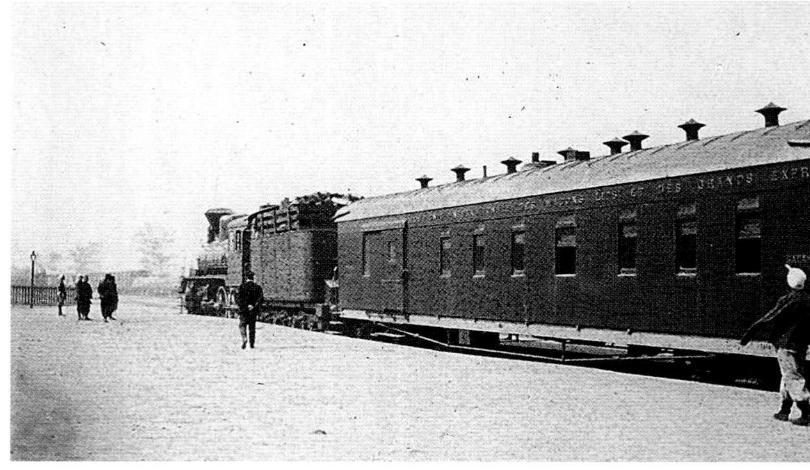

at the glorious mosques, the turquoise blues of the 14th-century tiles and the avenues of silver poplars against the backdrop of snow mountains. On arriving at Tamerlane's tomb, Ovitch told her that they would not be able to enter without a police permit. '"Nonsense, you will go in with my card",' she replied 'which he did, and we saw everything'.[71]

Ella had less luck at the Chinese border, where she discovered she could go no further than Osh as she lacked the necessary permits. So she returned by train and carriage to Tashkent, then took a train to Orenburg on the Ural river and drove on to Samara, from where she sailed up the Volga river to Moscow. She found Moscow badly organised, filthy and not much fun, although she enjoyed seeing the treasures of the Kremlin. She also lost the services of Ovitch, who was kicked out for being Jewish. He was ordered to leave within 12 hours, although she was hardly sad to see him go as she thought him a hopeless guide. Once

again, she was dressed almost in rags by the end of the journey and she asked Alice to bring some clothes for her to Paris, where they'd arranged to meet. From there, the sisters made their way back to Scotland.

In March 1912, soon to turn 51, Ella embarked on her second Central Asian trip with the intention of reaching Khiva. In St Petersburg, her Russian friends were incredulous that she wanted to go and the British Consul-General even suggested she might be a spy. An Embassy under-secretary was insistent that she must avoid any possibility of getting them all into trouble.

Ella struggled to find a guide. Eventually, she settled on an individual she referred to as Fritz, who worked at the Greek embassy and came with recommendations. Having acquired permits for them both and done the tourist trail around St Petersburg, she set off on the five-day journey to Tashkent, travelling second class because General Sukhomlinov, the Russian Minister for War, was also aboard the train and

below Ella then took a military steamer down the Oxus river from Charjui to the landing point for Khiva. Her fellow passengers included a general, two Mennonites and a couple of commercial travellers.

had taken all the first-class accommodation. At every station a red carpet was rolled out for the general and bands played, so there was a great deal of pomp to watch and enjoy. Ella thought Tashkent relatively civilised. When she was not exploring the old city, which was her principal interest, she spent time with her fellow Europeans, who formed quite a large expat community. Then she presented her Russian friends with tartan pin cushions as farewell presents and took an overnight train to Samarkand. There she returned to the shops she found so irresistible, picked up more

souvenirs and was greeted by the merchants, who recognised her from her previous trip. She travelled on to Bokhara, where a Russian diplomatic agent helped her to work out the best way of getting to Khiva. The usual option, which did not appeal, was to ride by camel. She then discovered that she could hitch a lift on a very basic Russian troopship from Charjui.

There was a terrible sandstorm as they sailed down the Oxus river and all the hatches had to be battened right down until it blew over; also, the boat kept going aground in the shallow water. Ella spent

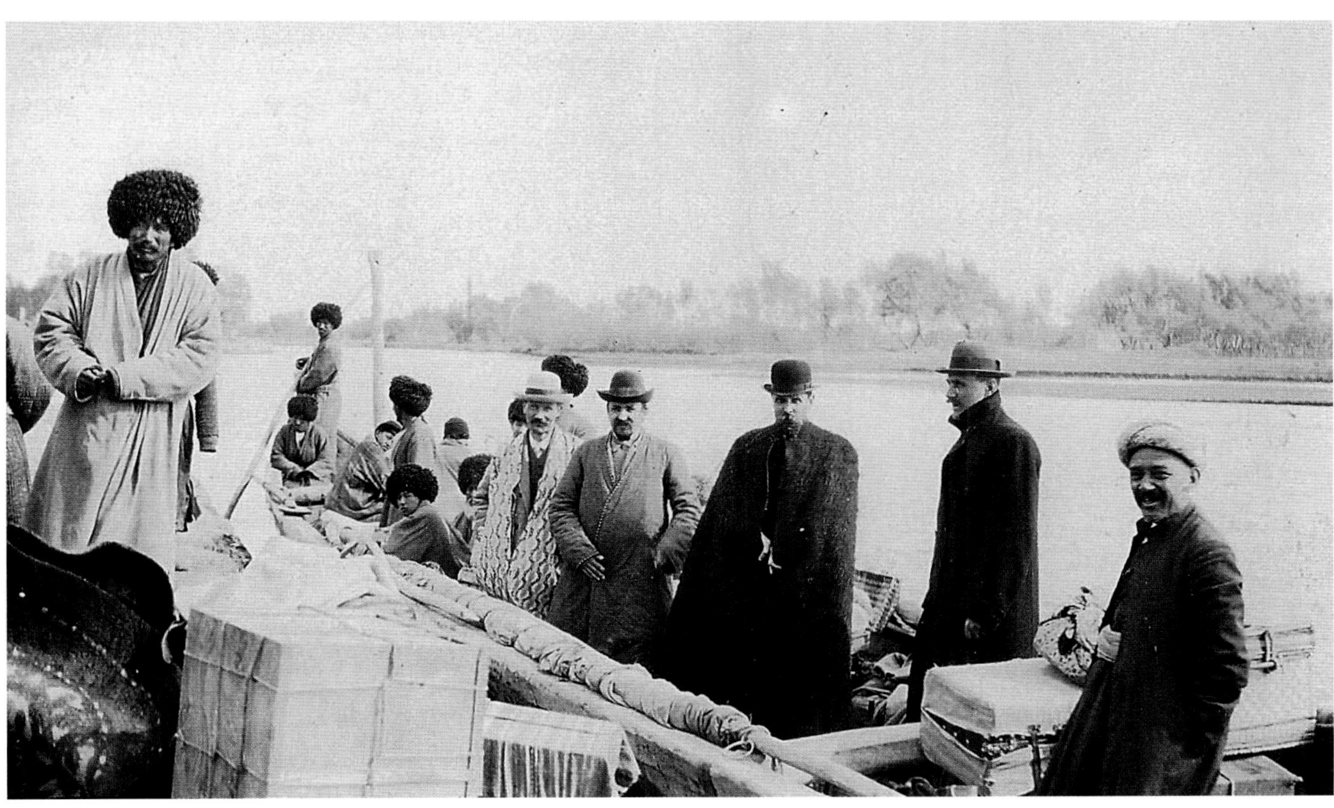

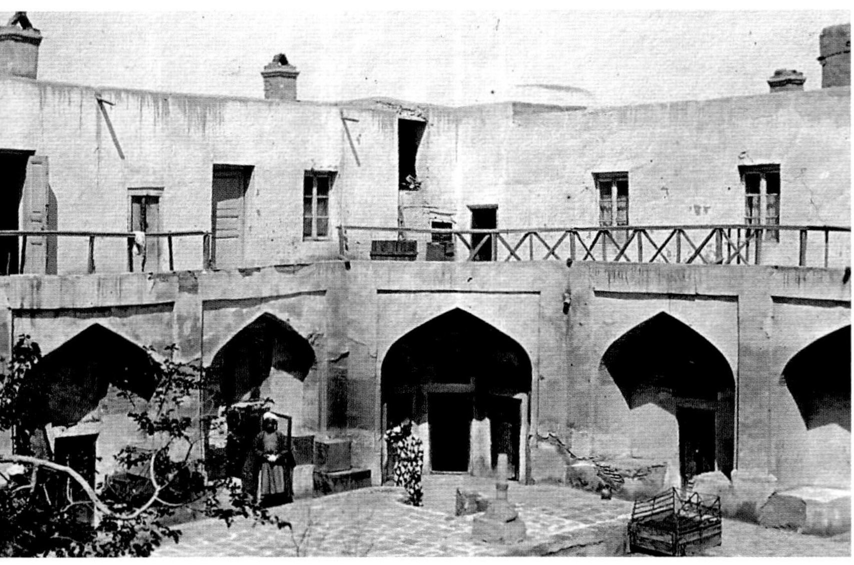

left A caravanserai in Bokhara, where Ella stayed. Her lodgings would have been room only; she had to provide her own camp bed and bedding.

much of her time on board conversing with General Sukhomlinoff in broken French. After less than a week, they arrived at a place with 'not a stick or even a stone to mark the landing on this blank shore'. The following morning, Ella boarded a *caiouque*. Sitting on her luggage on rugs spread over the deck, she continued her journey along the river, paddled by a Turcoman crew wearing dark red *khalats* and black sheepskin busbies. After some hours the boat entered a canal and the final miles to Novo Urgentsh were completed in a cart, or *arbas*, with enormous, iron-studded wheels. Ella spent several days in the town, with its partially destroyed mud walls, 'mud-built houses, with a raised platform in front, upon which the owner spends the greater part of the day, either for pleasure or business,' and cotton dealers. She then took a *droshky* – 'literally a basket on wheels' – to complete the final four hours across rain-freshened desert strips and oases, 'lines of green and gold, and gold and green, beyond which the walls and minarets of Khiva appeared ... all past difficulties and discomforts were forgotten'.[72]

A photograph of Ella, taken by a guard at the city gates, commemorates the fact that she was the first British woman to visit Khiva. Walking around unveiled where all the women wore a heavy black horsehair *chedra* must have made her even more of a curiosity. Most visitors were billeted at the Khan's palace, but this was not deemed suitable for a single woman. However, with her usual luck, Ella had been given a letter of introduction to a Colonel Korniloff, the resident Russian Commissioner and adviser to the Khan, and he invited her to stay. Korniloff and his wife Natalia Anatolia lived in a small but comfortable house that boasted such comforts as a steam bath. Ella had become quite flavoursome by this stage of her journey, so the bath was a great boon, although she struggled with the Russian tradition of communal bathing. She felt she had not known the family long enough to share their *bad stube* with them and eventually managed to convey this, although her hostess felt duty bound to superintend her ablutions. 'Suitable bathing attire not forming part of my travelling kit, my mind wavered between a nightgown or a Burberry, and finally the latter carried the day,' wrote Ella, noting that there was not even a piece of string to fasten the door of the bathing shed. Its stove had been kept burning and Madame Korniloff 'poured volumes of water on the floor to raise the steam ... the good lady was induced at last to leave me, though she twice returned to see how I was getting on'.[73]

Tucked up in bed with a glass of tea stirred with cherry jam, Ella slept peacefully after her day of novel experiences. She would later enjoy sitting on the veranda 'listening to a chorus of birds, too delightful. No rest cure could be more complete here. No letters come or go and we might be the sole inhabitants of the globe'.[74] The Korniloffs were generous hosts and guides, arranging for a cavalcade of *droshkys* to take her sight-seeing. Although her hostess spoke only the Khivan dialect and Russian, of which Ella had only a few words, they became great friends – Ella never let a language barrier stop her from communicating. She enjoyed an audience with the Khan, during which they exchanged signed photographs, and went on an excursion to a peaceful Mennonite colony, a member of which, Otto Toews, she had met on the Russian troopship. She was invited to stay and even shown a cottage as an inducement – 'they have a school and no teacher, so I was sorry I could not stay to help them, but felt I should never made a good Mennonite'.[75]

When she finally departed Khiva, a tearful Madame Korniloff pressed upon her a bag of camel wool and a box of pickled cabbage to sustain her on her journey. Ella never forgot the couple's kindness and, treasured among her souvenirs, she kept a photograph of them sitting on their veranda. The reverse bears the tragic postscript: 'murdered by the Bolsheviks during the October Revolution, 1916'.

Ella retraced her steps to Charjui and along the Oxus river to Tashkent, making her way by train to St. Petersburg and from there back to London. In 1913, she was made one of the first female fellows of the Royal Geographical Society.

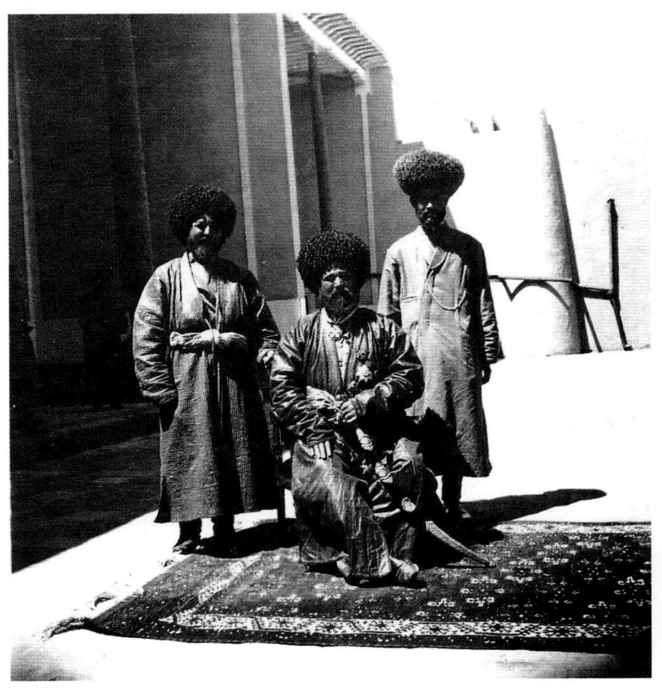

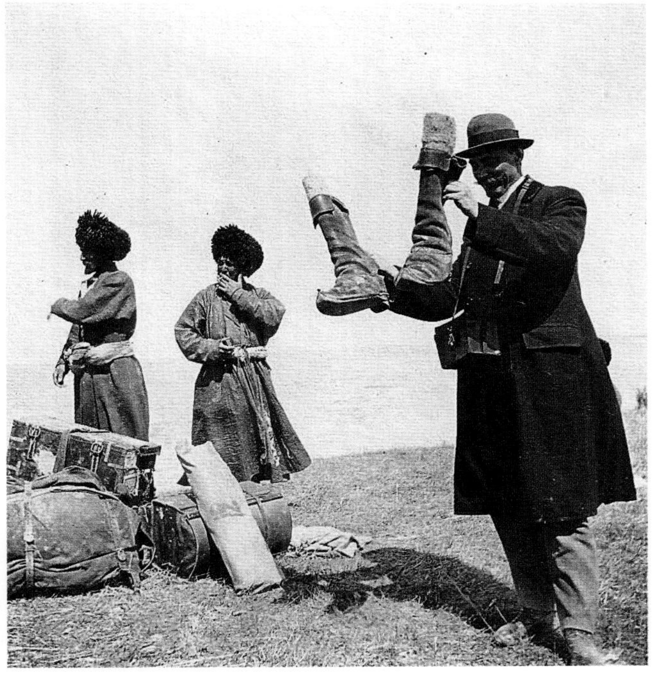

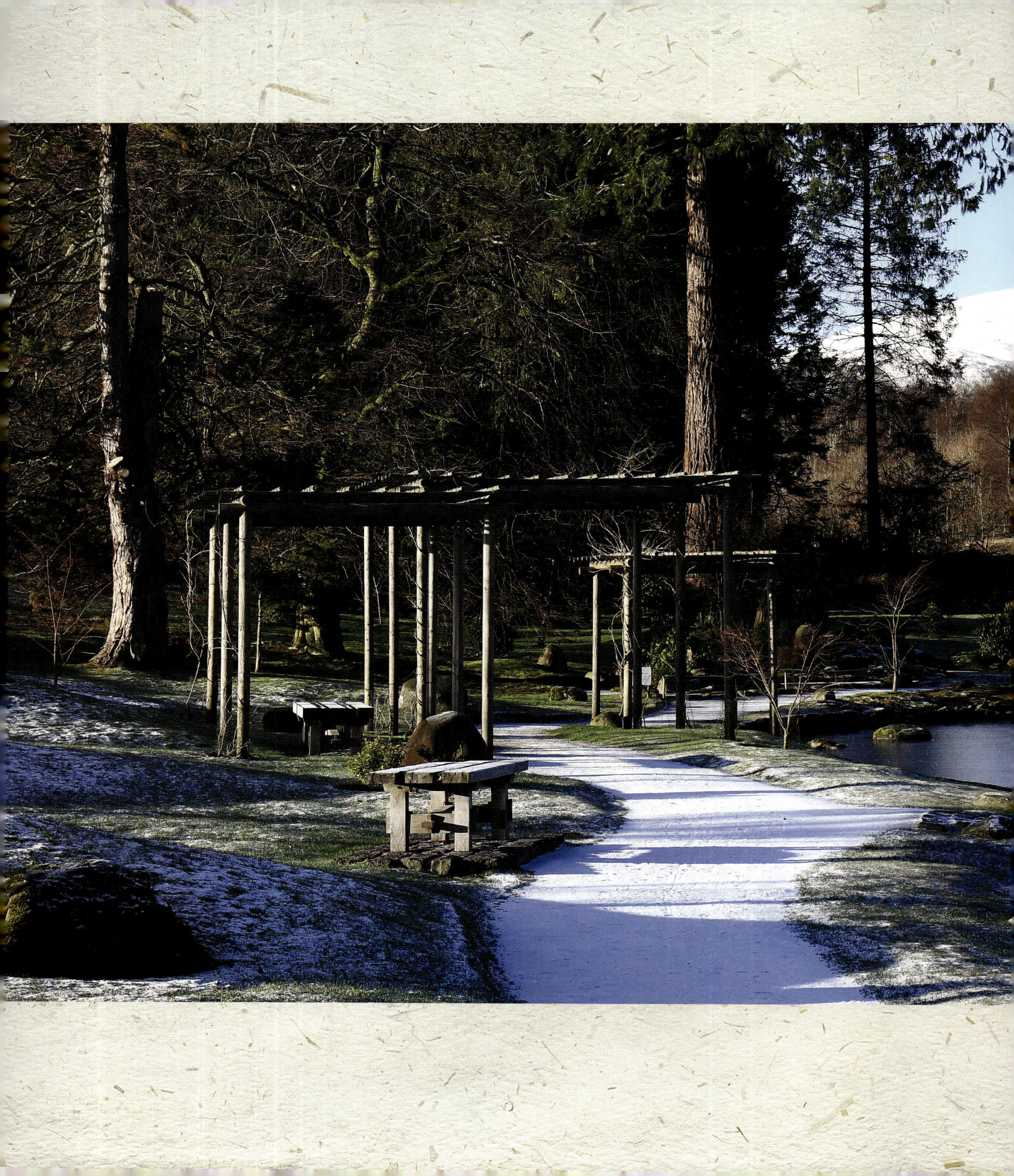

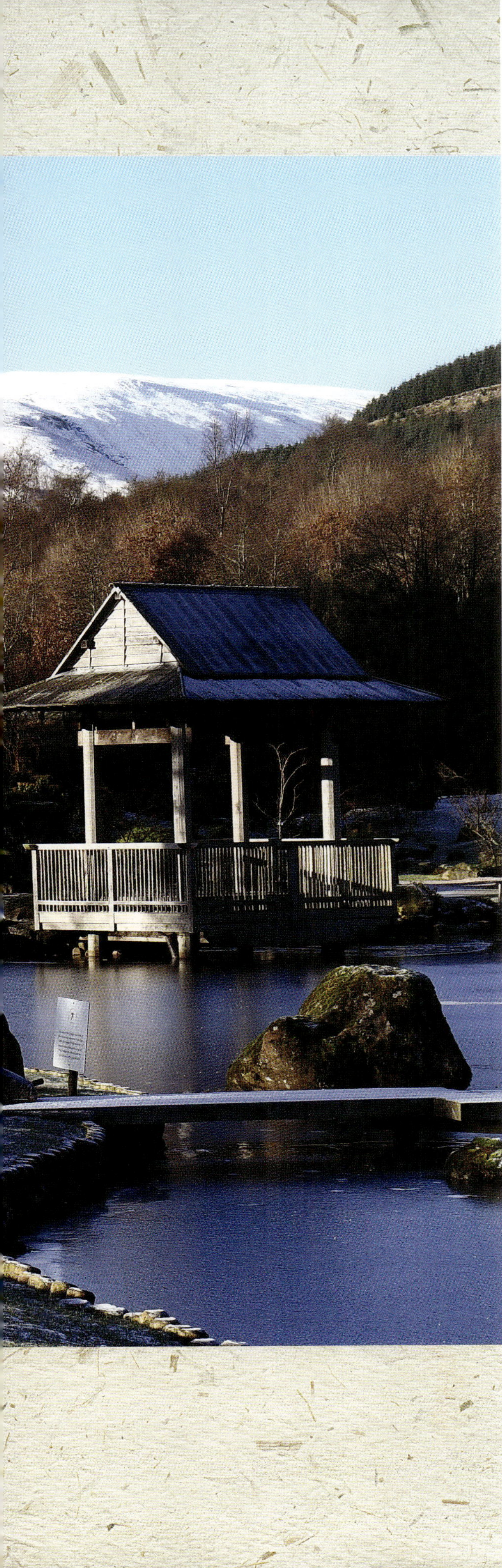

7 · America, 1914

ELLA'S FINAL BIG trip was very civilised compared to her previous ones, but the voyage didn't start well. Modern stabilising technology hadn't yet arrived and the Atlantic crossing on SS *Carmania* was the worst boat journey she had ever endured. The waves were enormous, leaving tables upturned and bodies tossed around. One desperately seasick passenger died and was buried at sea. For Ella, 'half an apple was all I could manage till this morning when I tried some porridge, and I could only get my face washed with the help of smelling salts. I consoled myself that if we were to go down I would do it as comfortably as possible under the circumstances ... Blyth [Abigail, her maid] is a frightful sailor and had to dash out of the cabin between the extraction of each hairpin. I smile as I compare this with my last boat on the Oxus, and my horror of the Atlantic has been fully realised'.[76]

In New York, Ella had the usual fistful of introductions and delighted in the hospitality she received. She met up with Franklin D. Roosevelt's sister and Mrs Alva Vanderbilt, mother of the Duchess of Marlborough, and, in Washington, took lunch with Mrs Eleanor Roosevelt – the future President was then Assistant Secretary to the Navy. She then continued

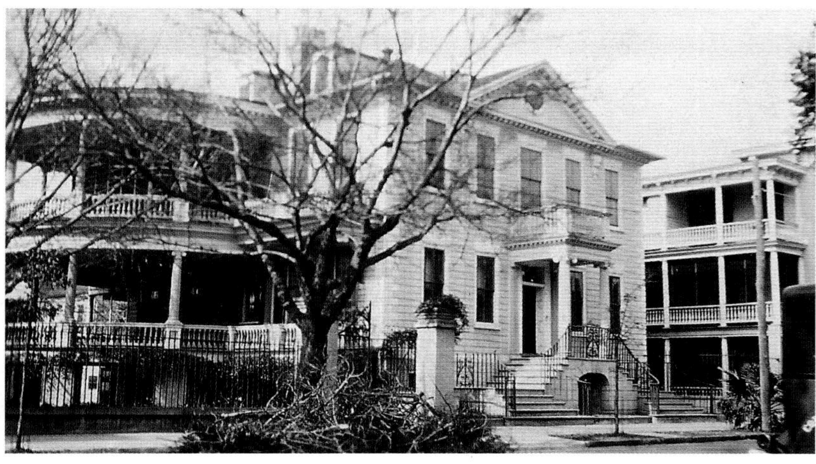

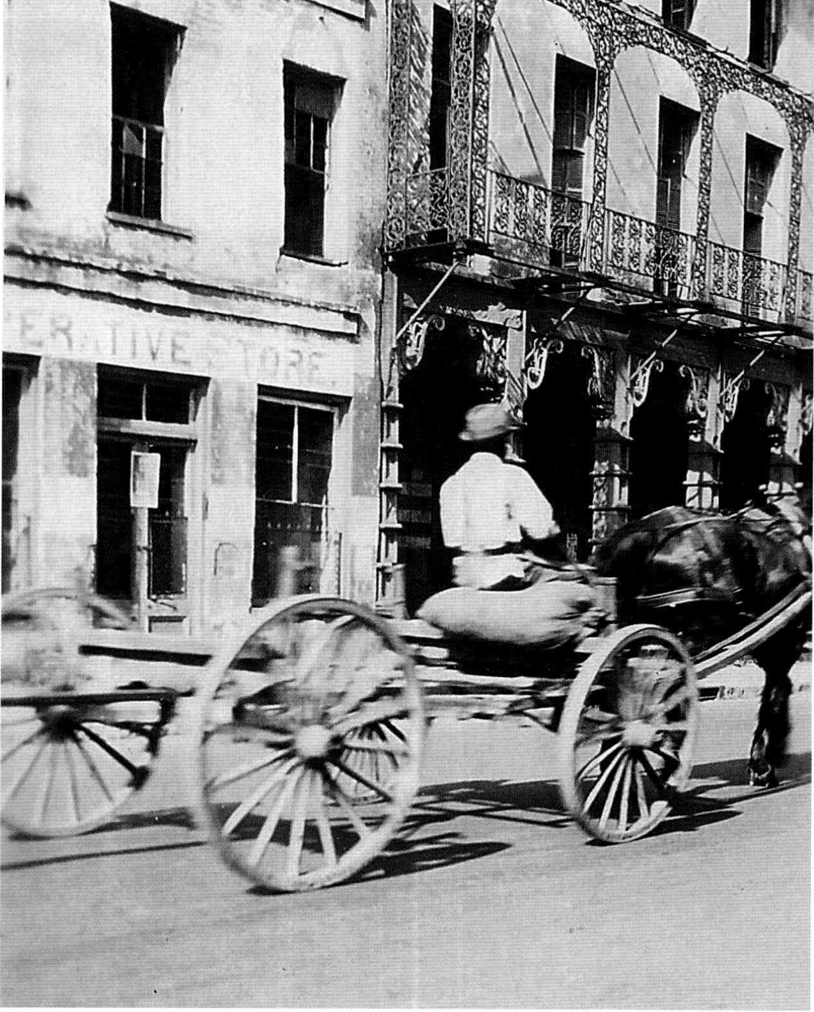

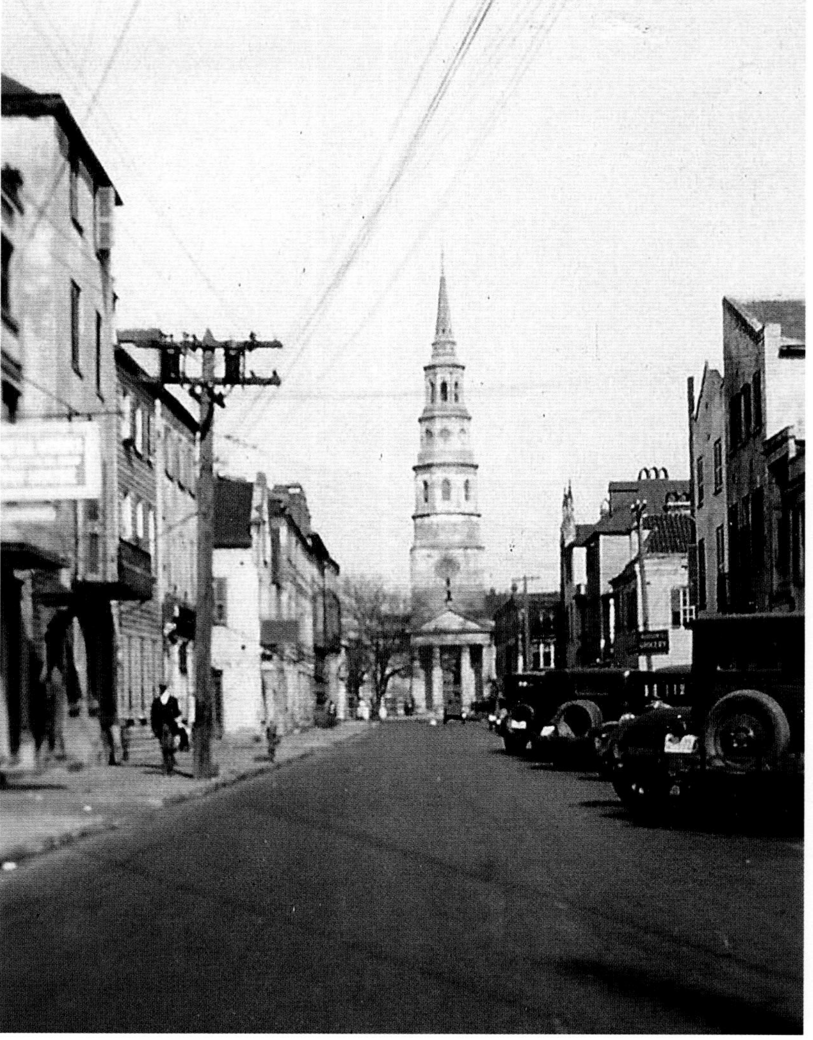

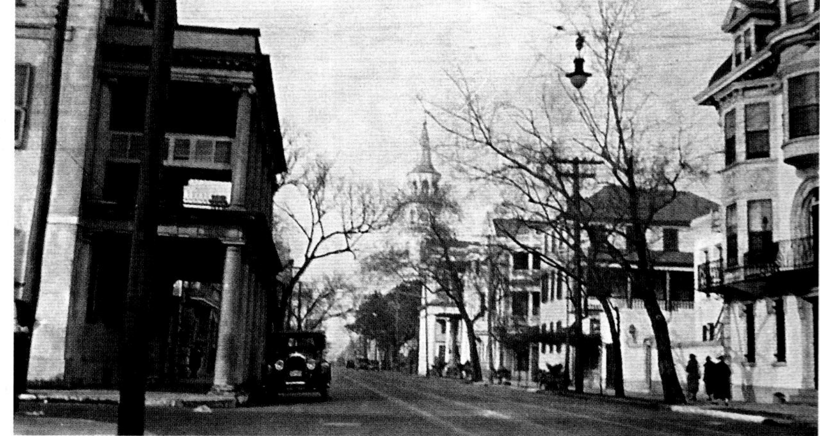

south to Charleston, where she found both the old houses and their inhabitants 'delightfully quaint,' their 'air of refinement' and 'gentility' a striking contrast to Cuba, where she arrived on March 25. 'I went to the Plaza Hotel and had a nice experience. I was just ready for bed when a shot rang out in the courtyard, and a frightened Blyth ran in to say that pellets had struck her window. I sent for the manager who was most rude and said he did not guarantee the safety of his guests.'[77]

After a 14-hour train journey to the Spanish town of Camaguey, where she and Blyth were devoured by ants and mosquitoes, but found the vegetation gorgeous, Ella returned via New Orleans. Here, they saw the Rotunda, the old slave market. 'It still has the auctioneers' bench with their names written above. One feels the tragic atmosphere in the covered shed where the slaves were kept.'[78]

Ella's great-aunt, Isabella (sister of her grandfather Alexander Christie), had left Scotland with her husband John Hill and settled in Baton Rouge, the capital city of the state of Louisiana, where he started a foundry. Ella stayed with their descendants at their Homestead Plantation in a house full of big old Scots mahogany furniture that reminded her of Cowden; she found the whole set up delightfully old-fashioned. A reception was thrown for her so that she could meet all the family and thereafter she stayed in touch with her American relations, who would send a big bag of pecan nuts to Cowden every year.

After the Grand Canyon and Los Angeles, Ella visited Santa Barbara, where she found herself 'in the country of millionaires, very different from the poor, proud South, and she found that local guides tended to describe each sight in terms of dollars'.[79] The next stage of the journey found Ella in Salt Lake City, where she had an introduction to another Scot, who had become a Mormon a few years previously and was keen to hear an accent from his homeland. 'As I was not sufficient in that line I turned Blyth on to him: "my, you're the genuine thing", was her reward.'[80] He introduced her to some 'brothers' of the Mormon community and showed her the Tabernacle.

Before returning home in June, Ella visited Boston and also the US Military Academy at West Point, where she attended the passing out parade of a distant cousin, John Hill Carruth. Ella was the only one of the young cadet's relatives able to go to the parade and, afterwards, 'John came to me at once and I was glad to be the one of his kin to congratulate him'.[81] John would go on to become a general, commanding US forces in Europe during the Second World War. On one occasion while in France, he flew from the battlefield to see Ella in Scotland, assuming that his by now octogenarian cousin would be at home. In fact she was staying in England, so he caught up with her there while she was in Windsor visiting her old friend Lord Gowrie, Lieutenant-Governor of the Castle. When they parted company, each man kissed Ella on the cheek; 'few ladies of my age could boast of being embraced by a VC and an American general *on the same day*'.[82]

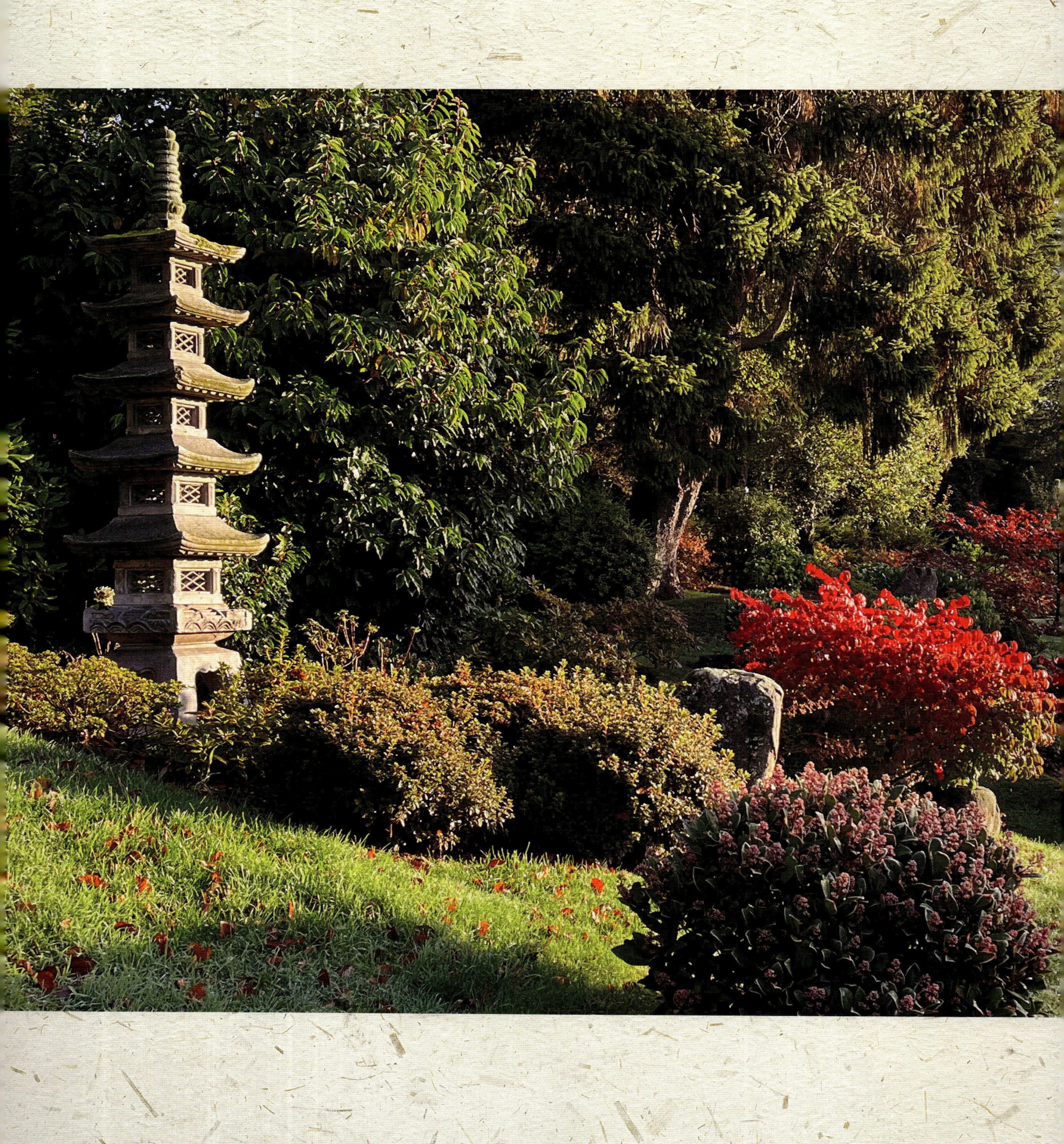

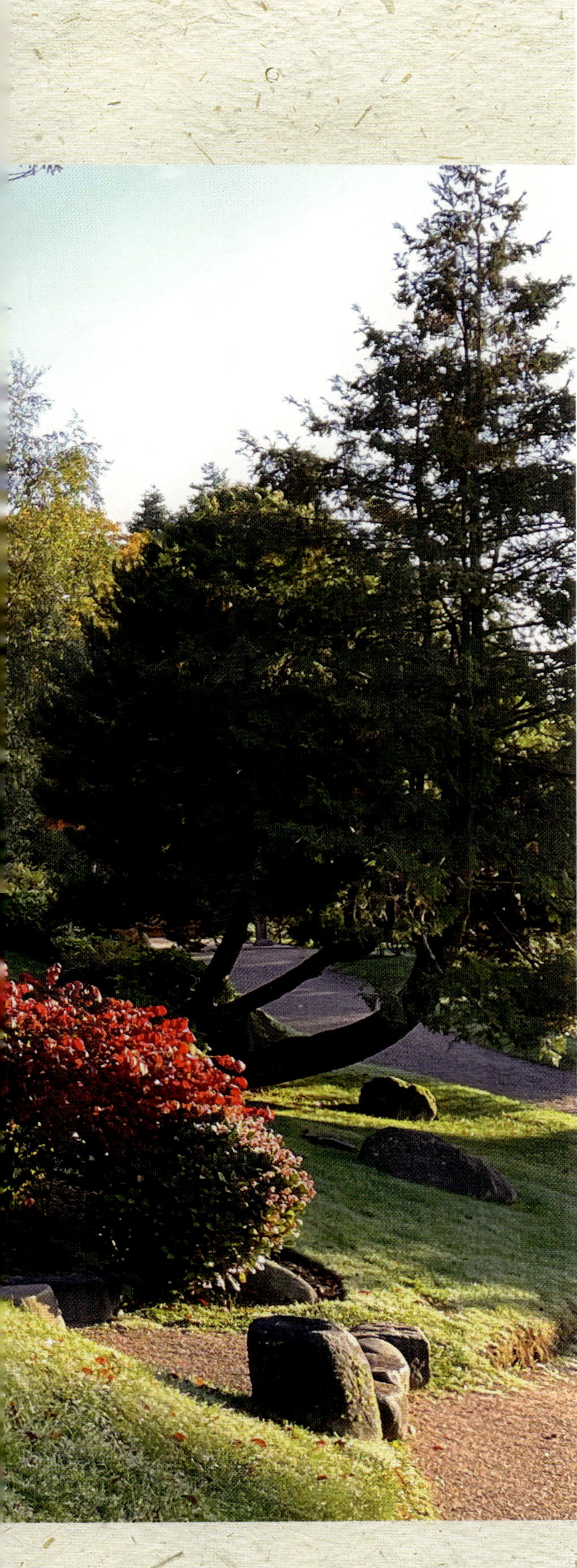

8 · The First World War and Later Years

ELLA WAS 53 when war was declared. There was no obvious role for her, but she raised money for military charities by giving lectures about her travels and did what she could to help her sister, who was president of the Lanarkshire branch the Red Cross. Alice and Robert began immediately to prepare for possible hardships. They ordered sacks of oatmeal and flour for each cottage on the Murdostoun estate and earmarked certain houses as hospitals, which Alice then set about staffing through her work with the Queen's Institute of District Nursing. Robert was involved with the Territorial Army and would be awarded a KBE for his war work. Their son Alexander, known as Bey, signed up with the Cameronians and would go on to receive a Military Cross and bar. Alice received an OBE, along with a citation from the British Red Cross Society praising her for being 'throughout the whole of the war an inspiration and a shining light to all who had the privilege of working with her'.[83]

In early 1916, the French Red Cross Committee in London asked Ella to go to France as manager of one of a network of cafes called L'Oeuvre de La Goutte de Café, officially known as Cantines des Dames Anglaises, which the French and the British Red Cross had established as places where soldiers could rest and relax. Ella arrived at Bar-sur-Aube with three fellow Scottish female assistants and asked Alice to send out much needed equipment and supplies, ranging from marmalade and cigarettes to sauces and chocolate. She described her daily routine in a letter to Elizabeth Haldane: 'The work begins at nine a.m. when our two orderlies have boiling water ready and we make coffee for four hundred. This is served at eleven after *déjeuner* with a cigarette each. Then we tidy up and prepare a dessert for the afternoon—custard or chocolate cream, something nourishing ... At three we have tea or coffee again, but coffee is always ready for any dropping in'.[84]

They were within earshot of the guns, which rattled the windows from 50 miles away. Many of the rations that Ella continuously requested from Alice failed to arrive, but the canteen provided a respite, albeit a fleeting one, from the horrors of the Battle of Verdun and Ella and her ladies acted as mothers to the French soldiers who turned up on their doorstep. 'We give each a glass of wine, a cigarette and a small souvenir, shake hands with as many as possible and wave them off as cheerfully as we can, when we think of what lies before them.'[85]

By October 1916, the fighting had intensified. 'One can never get over the stricken look in their eyes, as each morning sees them go off to take their part in a veritable hell,' Ella wrote. 'Very heavy air today, which gives one a headache. The gas fumes come over, and we have often noticed this after heavy fighting'.[86] The four Scots women continued to dispense their

cheer until February 1917, when they were sent back to Scotland. The strain had been immense, but after a period back at Cowden, Ella returned to France to run another canteen, this time at Mulhouse in Alsace, from 1918 to 1919.

Ella picked up the threads of her pre-war life and became active again in the Royal Geographical Society, the Central Asian Society, the Women's Guild and the British Red Cross Society. She was also a fellow of the Royal Geographical Society of Scotland (vice-president from 1934) and of the Society of Antiquaries of Scotland. She attended London concerts and talks by leading authors and made annual visits to her Parisian seamstresses. She also resumed her travels; her passport, issued in 1926, contains stamps for Egypt, Turkey, Russia, Cyprus and Germany.

But much of her time was spent at Cowden, where running the big house and estate was almost a full-time job. Many old friends came to stay, as well as new ones whom she'd met on her travels – she corresponded with acquaintances all over the world, sending them shortbread, books, woollens or Pond's Extract as presents. For her guests, she organised outings and teas in the Japanese garden, which she continued to develop with her gardener Shinzaburo Matsuo.

In 1927, while Ella was in America, her nephew Bey died as the result of a minor condition caused by a shrapnel injury in the First World War, leaving a young widow and two small children, Grizel and Robert. Ella rushed back from her holiday to support Alice, who had already lost one son (Robin, who died aged six) and would lose her husband three years later. The two

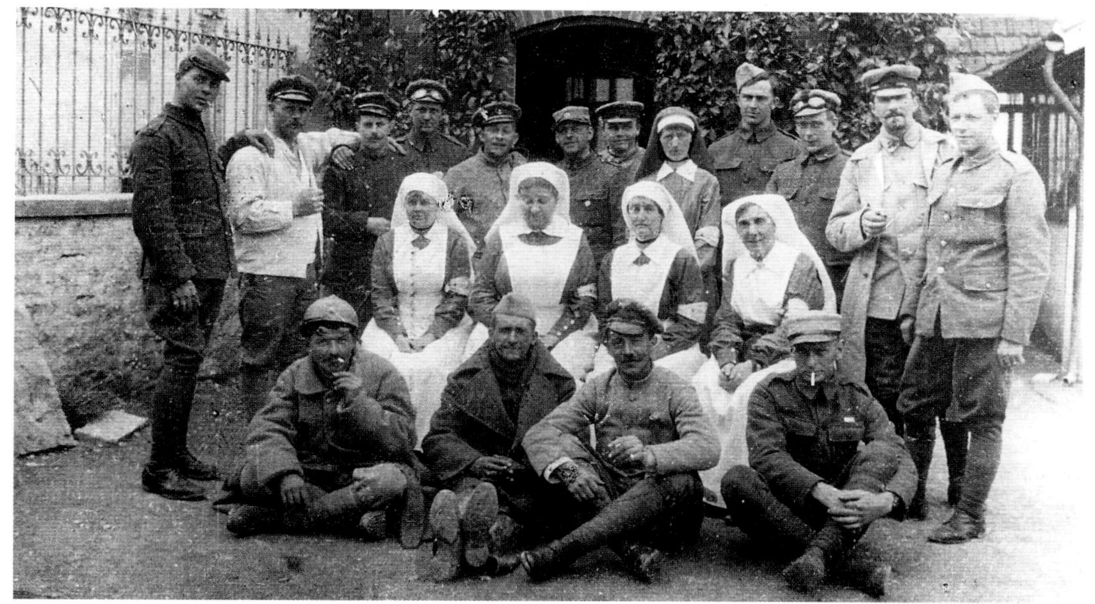

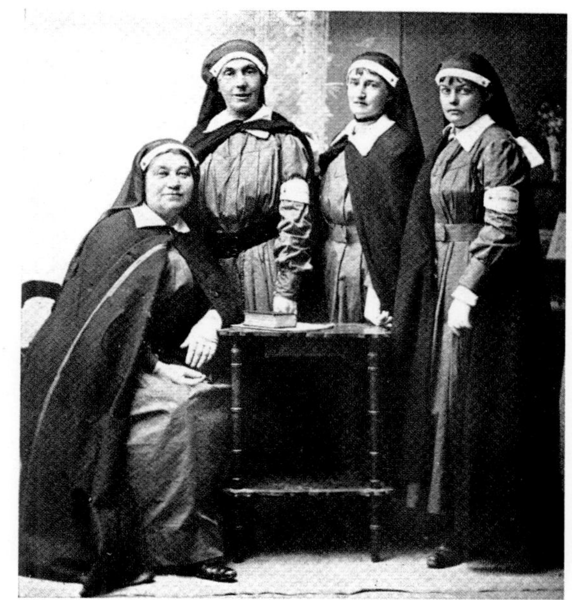

sisters became close companions again. They made trips to the continent together, staying at spas within easy reach of art galleries, concert halls and other attractions, and wrote their joint memoir, *A Long Look at Life By Two Victorians*, which was published in 1940. To the end, they were lively conversationalists, always full of interesting things to say. Alice's sole surviving son, John Stewart, who inherited Murdostoun, married Violet Douglas Home (known as Averil or Ava). 'When I first met them,' Violet would later write in the introduction to her book *Alicella*, 'they were nearing their sixties, but, as a friend has well said: 'one never thought of them as young or old, they were any age – or none.' Singly, each was a delightful companion; together they were a party. With Ella it seemed always holiday time, even when one was busy. To go abroad with her was to watch the world through her all-seeing eyes'.[87]

In 1940, after a decade of widowhood, Alice caught a chill while working in her garden and died aged 77. Ella was now left on her own – except that she was never entirely alone, and never allowed her spirits to dim or her enthusiasms to waver. 'As I glance round the rooms at Cowden,' she wrote in *A Long Look at Life*, 'my eye lights on article after article – coins, china, beads, bells, bracelets, and much else – serving to remind me of a score of journeys about the world.'[88] She wrote her third cookery book, with tips on how to use wartime rations as efficiently as possible, and continued to attend auction sales, buying porcelain and other objects for the overflowing rooms at Cowden Castle. 'It's not what you buy, Ma'am, that you regret – but what you *don't*,' she advised the keen royal collector, Queen Mary, who visited Cowden in 1937.[89] Believing that it's never too late to learn something or to join a club, Ella enrolled as a life member of the Kinross-shire Bee Keepers' Association at the age of 83.

As she grew less mobile, she would sit reading with a dog happily asleep on her lap, visited regularly by her great-nephew Robert (Bobby) Stewart and his sister Grizel, who lived at Arndean across the valley, and by friends who loved to drop by for a chat. Ella was suffering from leukemia when, aged 87, she died at Strathearn Road, Edinburgh on January 29, 1949. She was buried in the family plot at Muckhart churchyard.

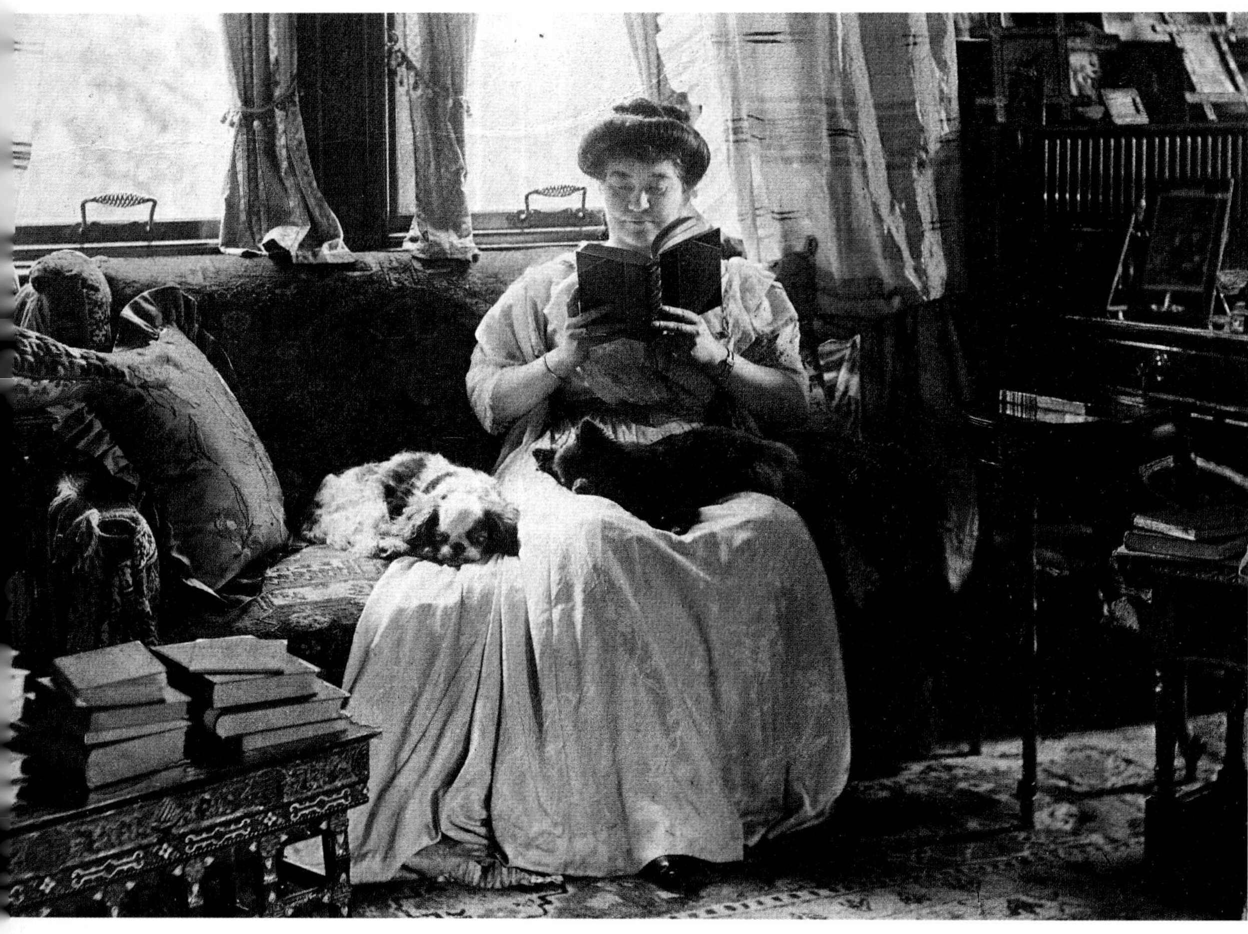

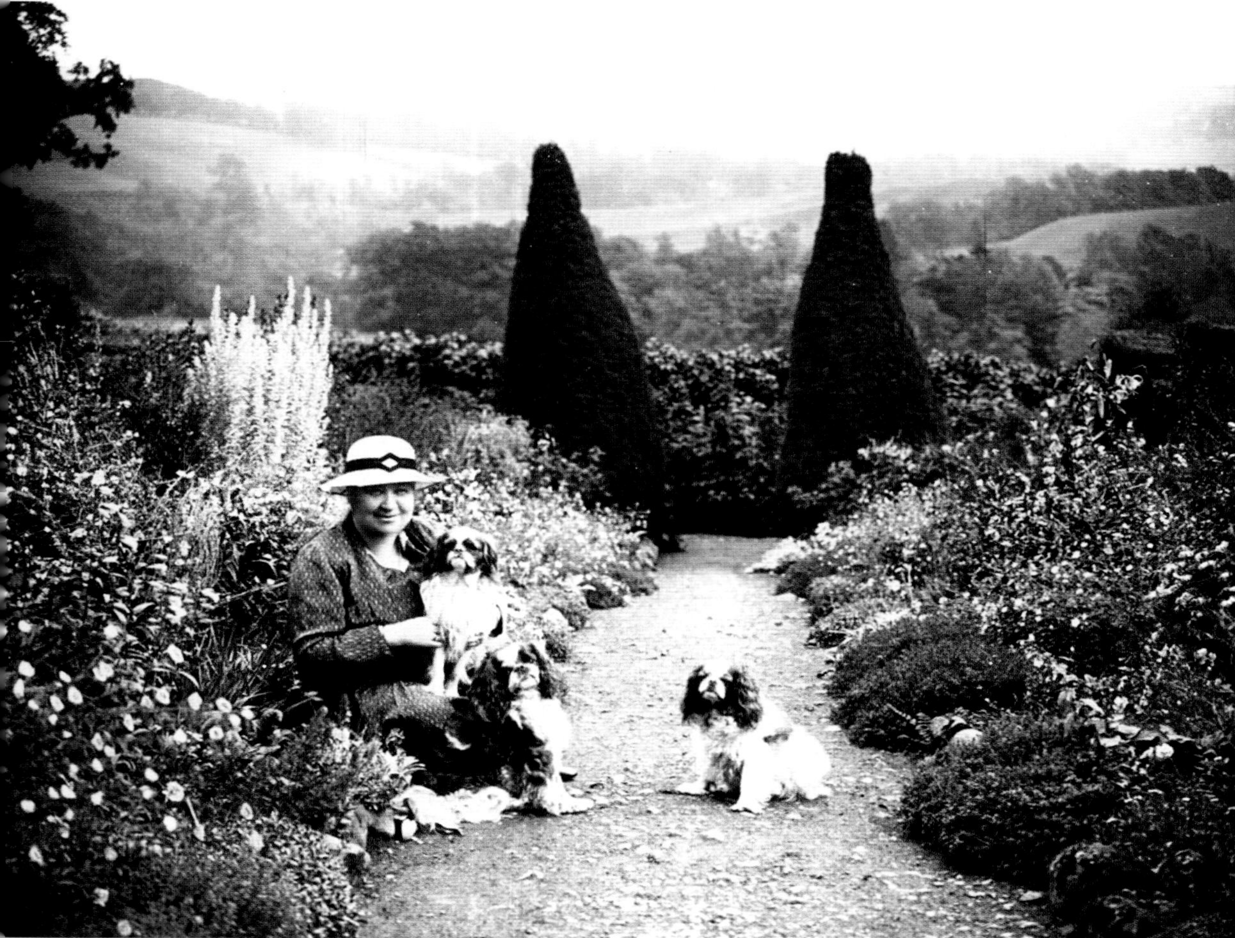

above Ella reading in the library at Cowden with her dogs.

below Ella and her dogs in the main garden at Cowden.

opposite above Ella sitting at the west end of the garden towards the end of her life, with the boathouse and fishing hut in the background. At her feet is what was called in her day the Lily Inlet.

opposite below left Looking east from the Master's Isle, with the North Gate on the left and the *taikobashi* in the middle distance.

opposite below right Ella and a friend on the second floor of the boathouse in the early 1930s. The *taikobashi* (just visible on the right) was changed by Professor Suzuki into a *yatsuhashi* – a low, zig-zag pontoon representing the twists and turns of life – in 1933.

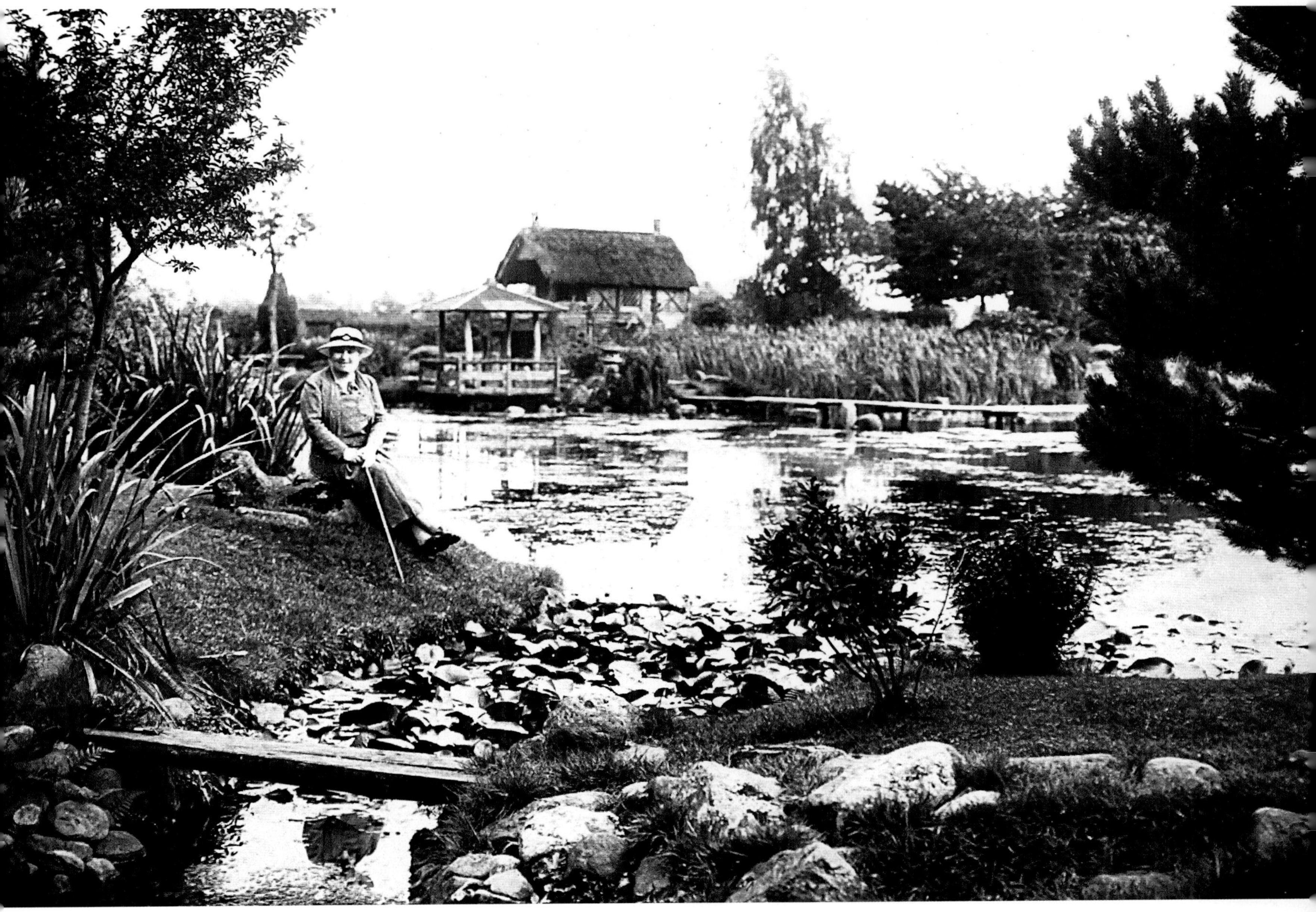

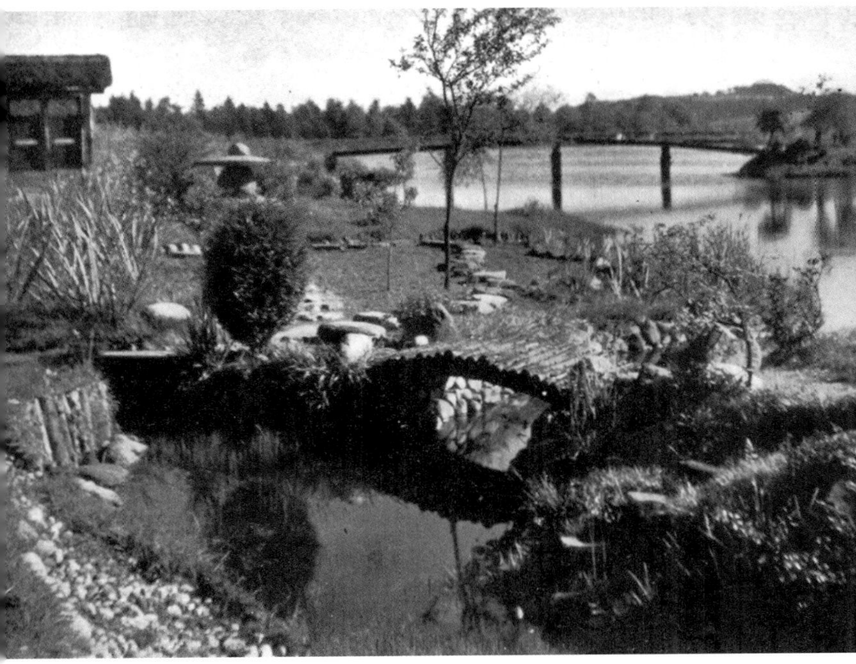

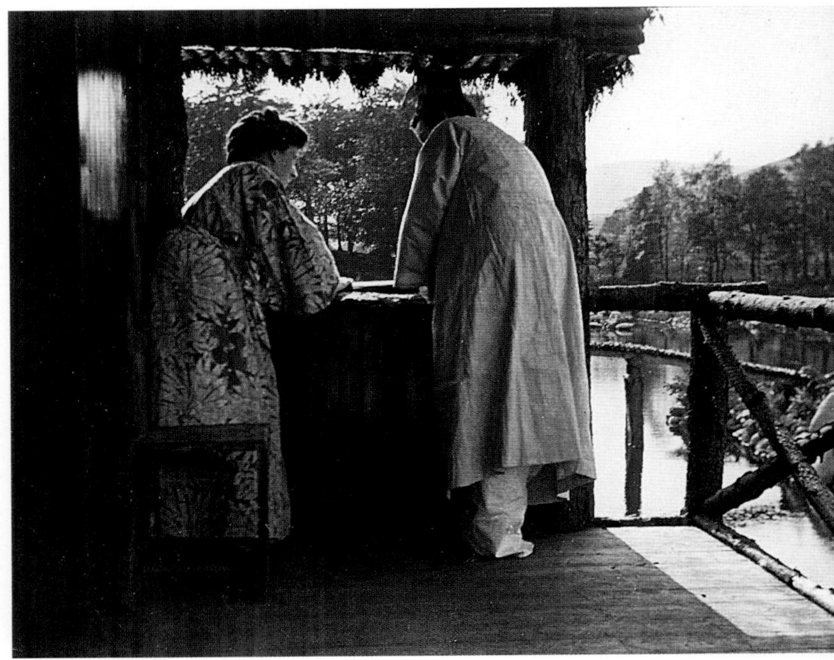

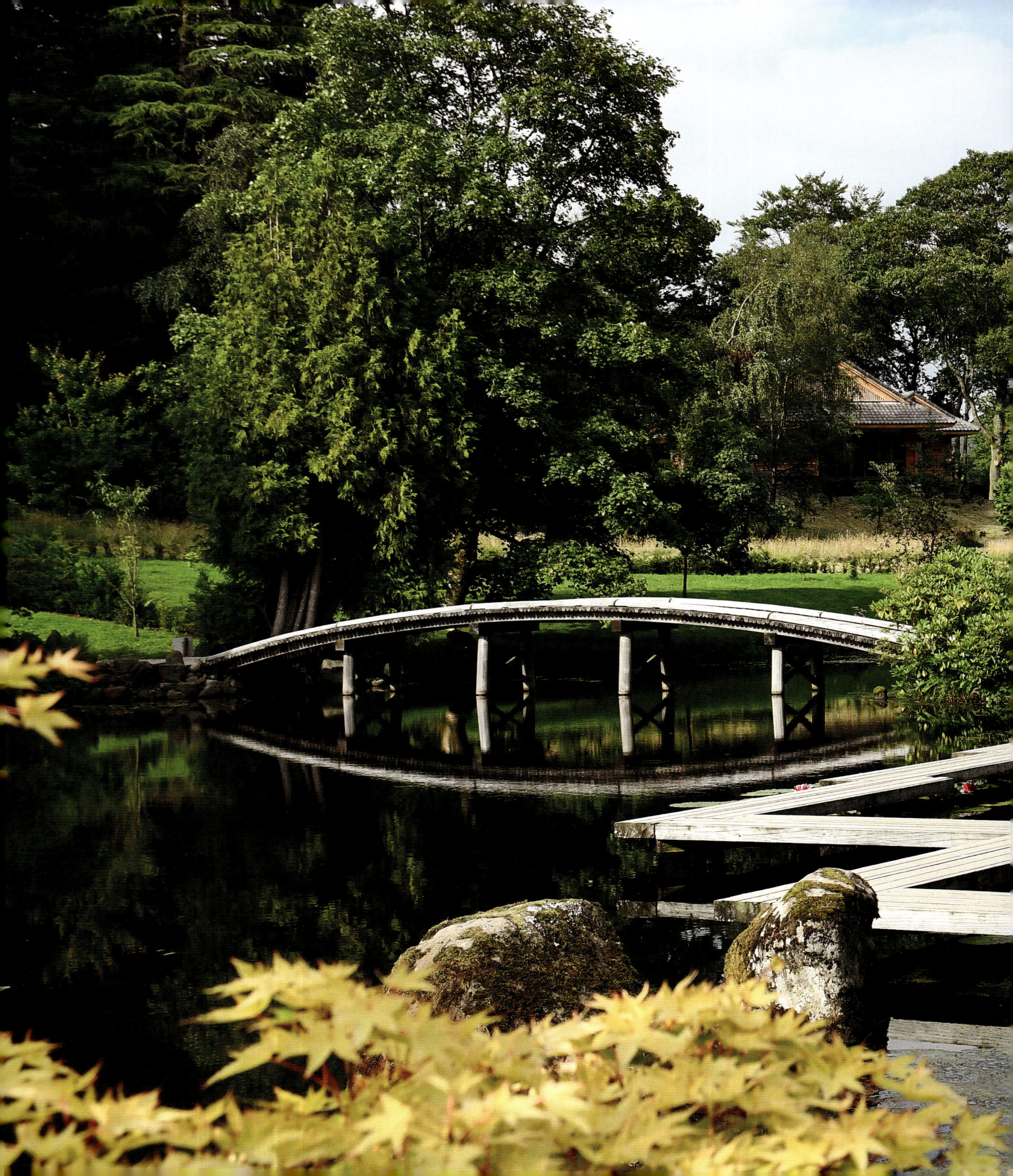

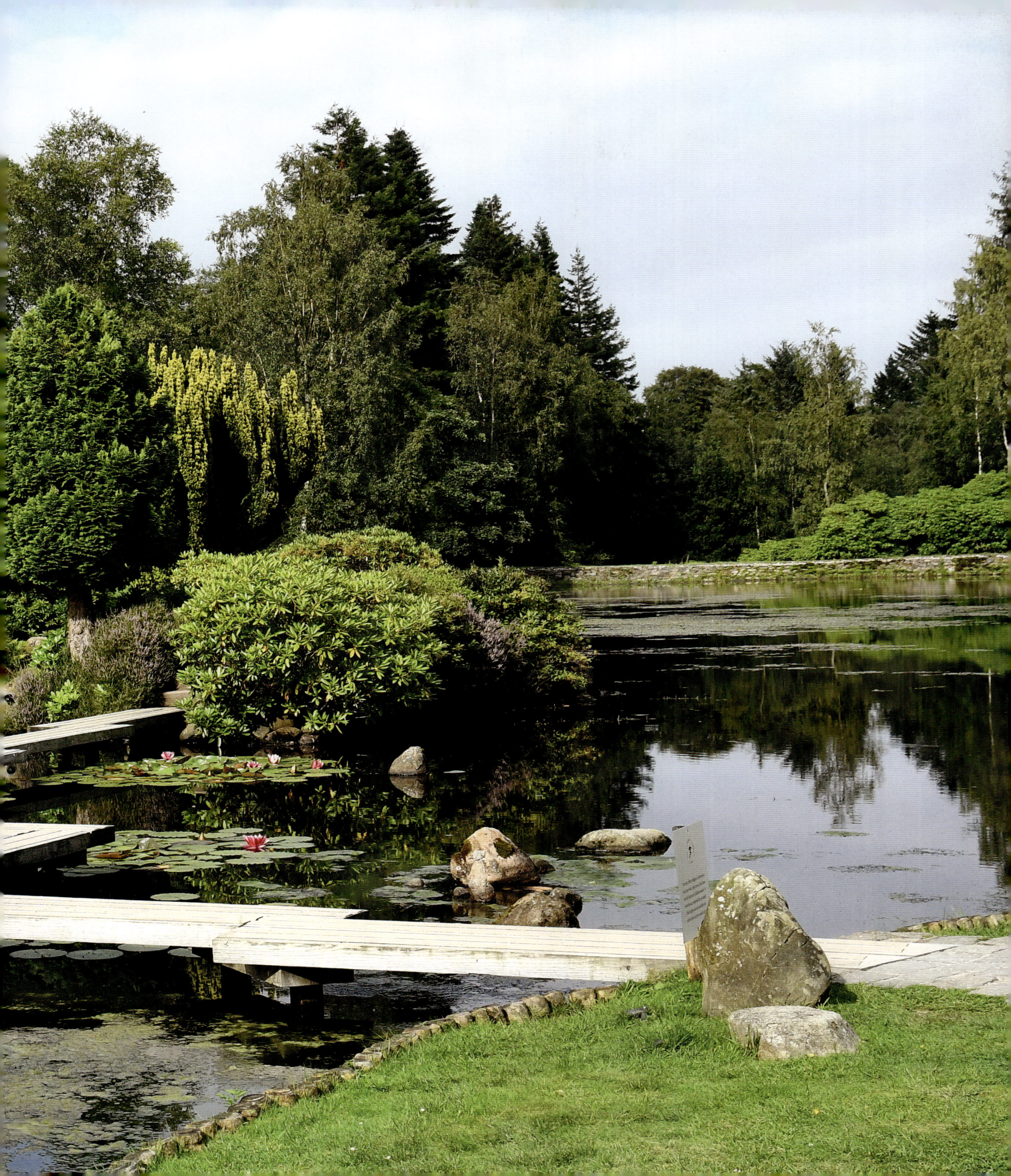

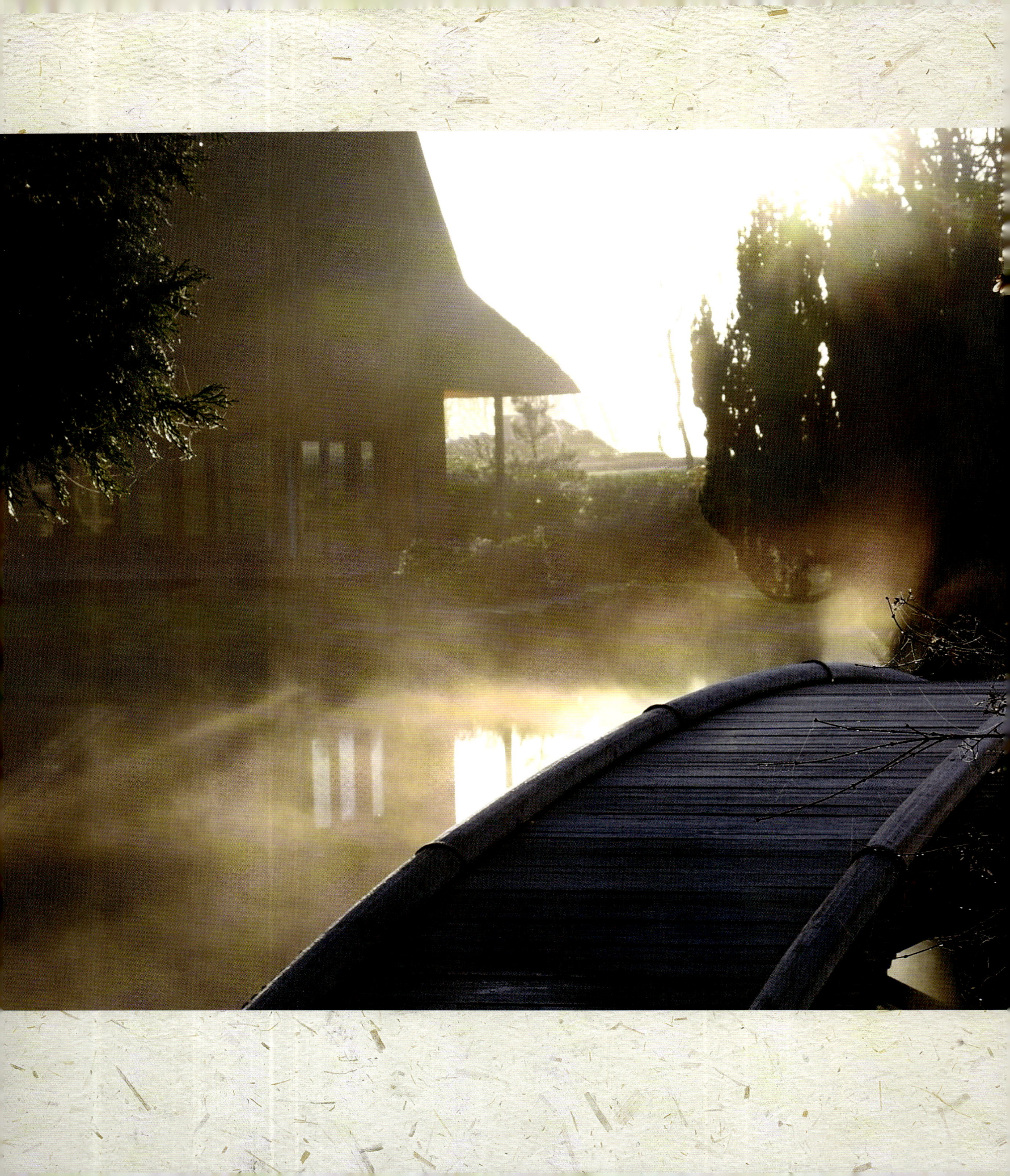

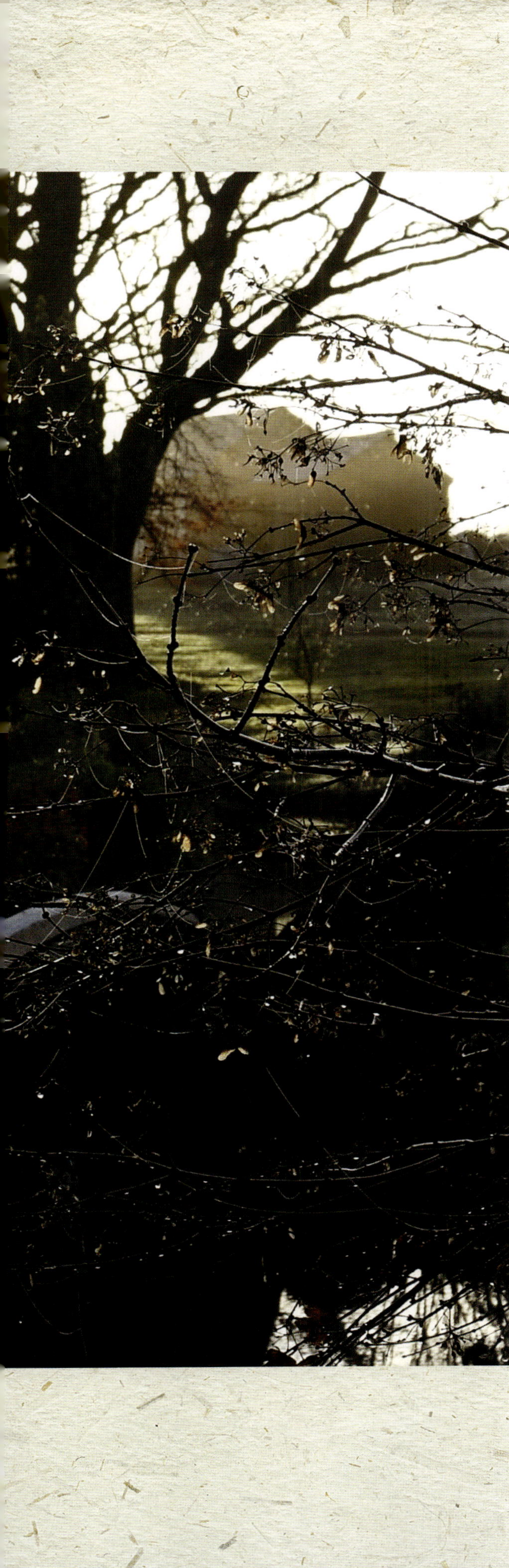

9 · The Japanese Garden

In a sheltered foothold of a grassy range of hills that stretch from sunrise to sunset lies the garden of my dreams. After the fierce volcanic agencies that upraised them, and the long eons of time that have moulded their undulating lines, the softly rounded hills encircle "Sharaku-En," "the place of pleasure and delight".[90]

THE JAPANESE GARDEN at Cowden is Ella's greatest legacy. Other British gardens featured Japanese elements, such as teahouses, curved red bridges and stepping stones, without adhering fully to the principles and planting of Japanese garden design, but at Cowden, Ella replicated what she had actually seen in Japan. Before telling the stories of how this garden came to be, of the people who helped to create it and its later development, we should consider the historical background.

There had been trading routes between Japan and the rest of the world since the 16th century, but they were tightly controlled and there was only one small entry point into the country for foreigners. This was a trading post on the artificial island of Dejima, off the shore at Nagasaki. The Portuguese were the first to be based there, followed by the Dutch. They were not allowed into Nagasaki, except with special dispensation from the local ruler, and Japanese citizens were only allowed to cross to Dejima if they were designated traders. Although cultural crossover was limited, occasionally foreigners did travel north to Edo (modern Tokyo). One, Englebert Kaempher, a botanist, physician and explorer, had an audience with the Shogun, during which he was quizzed extensively about how the rest of the world was developing. Then, in 1853, Commodore Matthew Perry of the US Navy arrived in Tokyo Bay with four ships, ostensibly to demand that shipwrecked whalers washed up on the shores of Hokkaido be treated more humanely, but actually with the aim of forcing Japan to open up to American trade. The following year, the US and the Shogunate signed the Treaty of Kanagawa, establishing an American consul on Japanese soil and the opening of two new ports.

In 1886, there was effectively a *coup d'etat*, when feudal lords in Kyoto ousted the last Shogun and reinstated the Emperor. This was the beginning of the Meiji Era, a time of enormous change in a country that had been isolated for 200 years. The Japanese began to participate in international exhibitions, such as the 1910 Japan-British Exhibition in London, to show their arts and customs to the world. Foreign experts known as *oyatoi* flocked to Japan to help develop buildings and transport infrastructure, and to teach the newly established government departments how things were done abroad. Many of these *oyatoi* were British, including Josiah Conder, an architect who helped to design buildings in Tokyo and who also had a great

interest in Japanese plants and gardening. He married a Japanese dance instructor and became totally assimilated into the country, where he would live for the rest of his life.

Conder's book, *Landscape Gardening in Japan* (1893), had a strong influence on the design at Cowden and Ella's copy is still in the family's possession. His interpretation was based very much on Meiji concepts, which saw a move away from the spiritual towards more aesthetic considerations in garden design. Many British Japanese gardens embraced the look with a profusion of lanterns and other ornamental features.

'I think the great secret of the Japanese talent for gardens lies in their superlative powers of imitation. The gardens were first copied from the Chinese, and then improved upon the lines of nature till one can scarcely see where the artificial and the natural join,' Ella reflected in a letter to Elizabeth Haldane written in Kyoto on April 23, 1907. The idea that she could do the same back at Cowden took root and, by May 3, she was writing to Alice: 'If feasible I shall bring a Jap home to lay out my pond. It could be made a dream of beauty'.[91]

There has been much academic discussion concerning the authenticity of Edwardian Japanese gardens, a vital question being whether or not they involved a genuine Japanese designer.

While in Tokyo, Ella started to put feelers out as to how she might find one. On May 9, Conder wrote: 'In reply to your letter of yesterday date I think you cannot do better than apply to the Yokohama Nursery Company at 21–35 Nakamura, Yokohama, otherwise

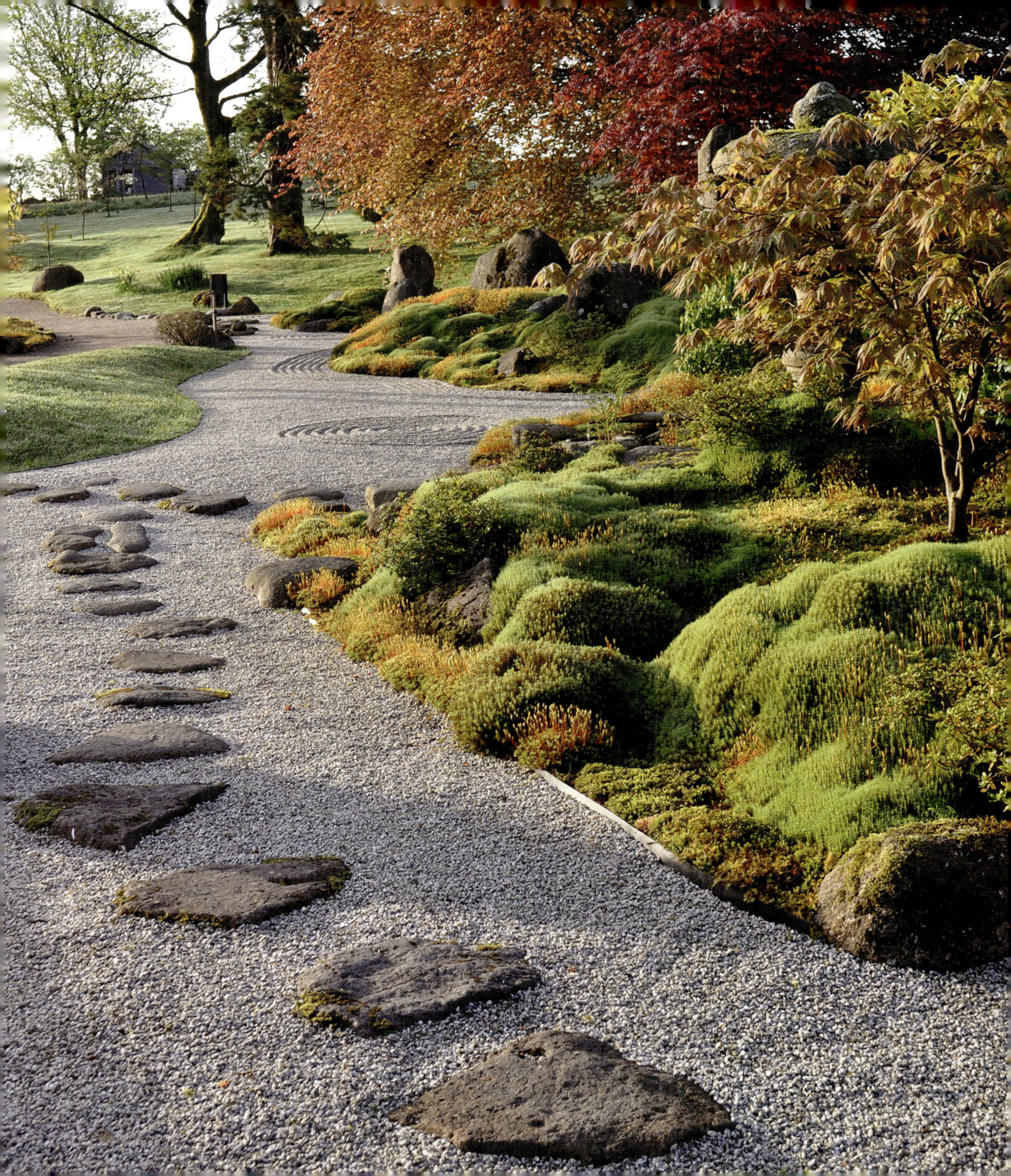

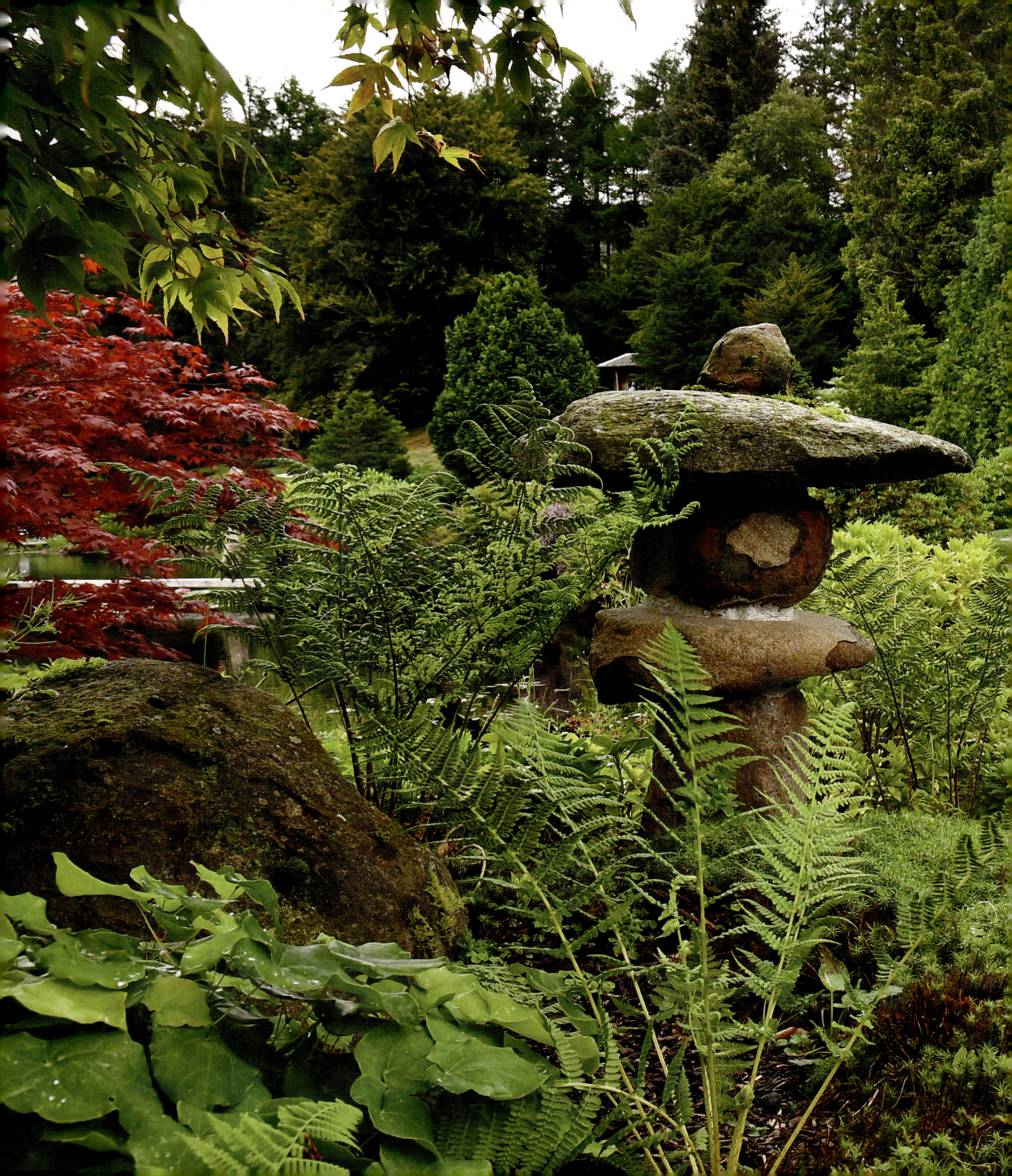

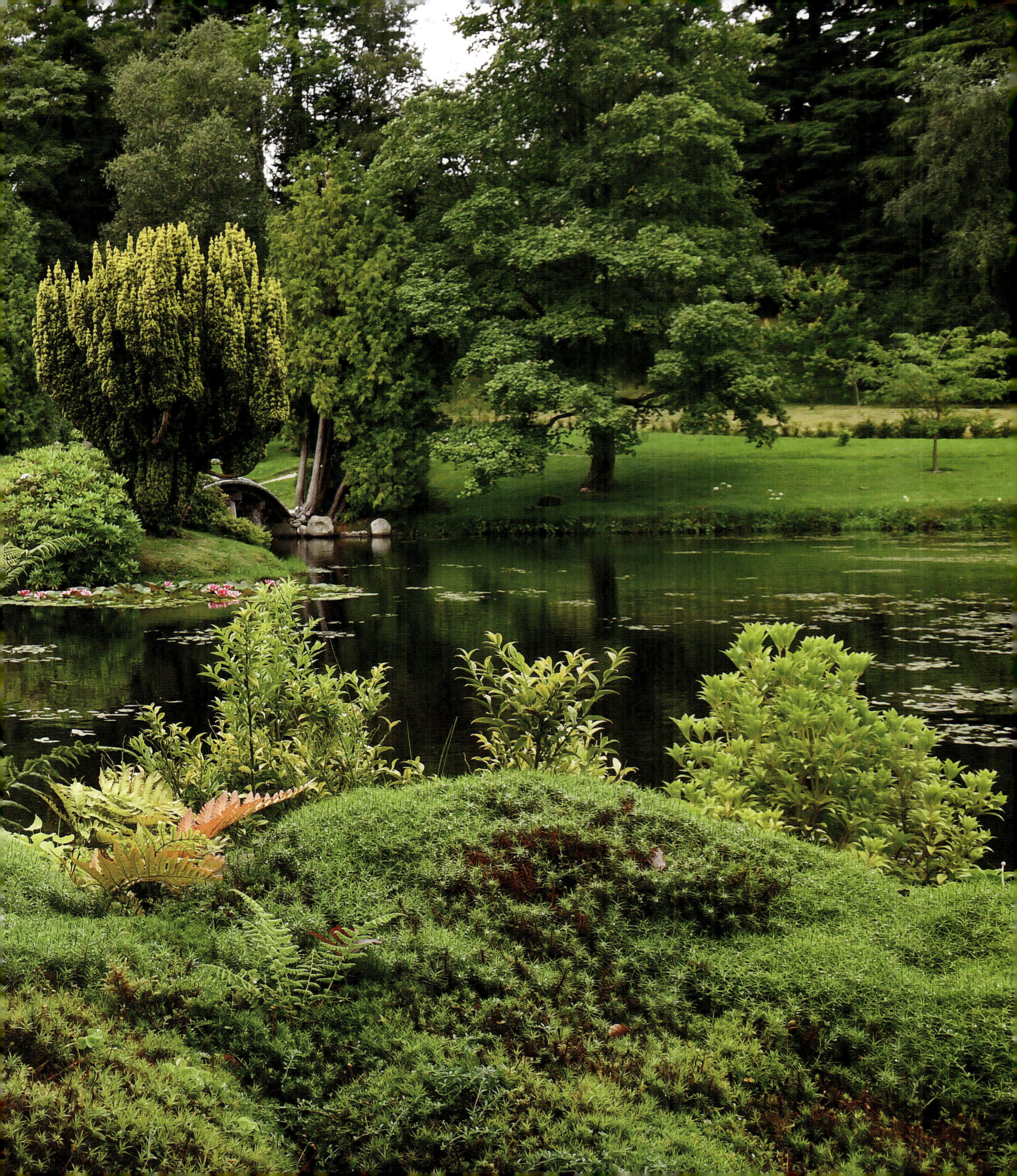

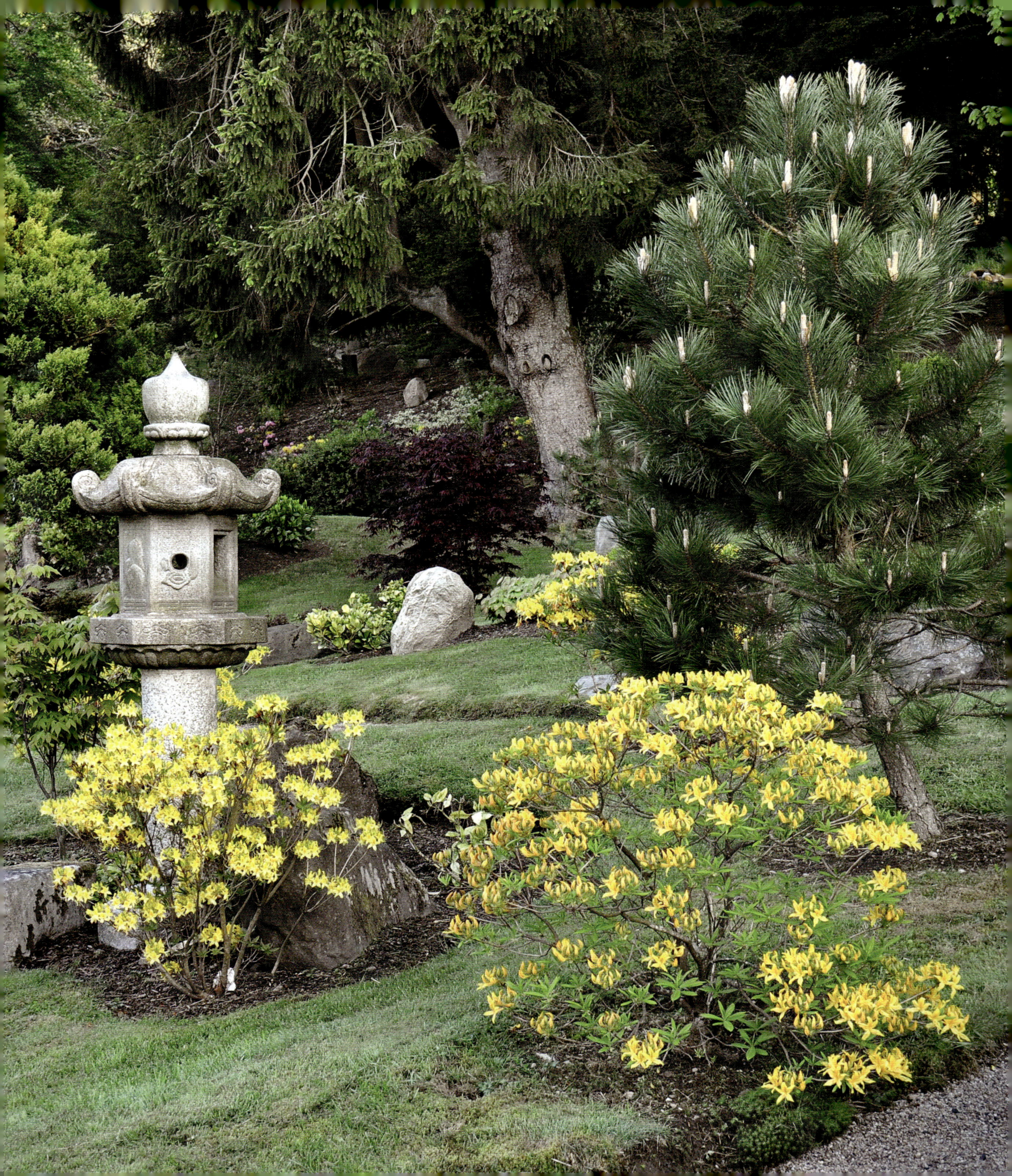

called the Gardeners Association. You will find there English-speaking experts acquainted with all branches of gardening who will if asked, I think, supply you with suitable Japanese designs for the garden as well as advising you upon selections of suitable stones, shrubs and lanterns etc. None of the Tokio [*sic*] gardeners would be able to assist you in the same way: they lack both organisation and acquaintance with matters of proper experts and attempt to get what you require from them; you would have to put yourself into the hands of guides or unreliable influencers and would in all probability be greatly disappointed'.[92]

Ella found the horticulturist Taki Handa (1871–1956) through a different route and, in 1907, commissioned her to design the garden at Cowden. Born in Kurume, Kyushu, the daughter of a prison guard, Handa had left school after her primary education in order to earn money to help support her family, which hailed from the samurai class but had fallen on hard times. When her brother qualified as a doctor, he insisted that Handa return to school, where she studied Chinese literature and weaving. She also took English lessons with the wife of the Anglican priest and, aged 16, converted to Christianity. In 1895, after training to be a teacher in Fukuoka, she went to Kyoto, where she studied botany, horticulture and English at Doshisha Women's College. She then moved to Tokyo, where she spent three years teaching at a missionary school. Back in Kyoto, she was introduced to the Le Boutillier family; she became their translator and helped them with plant collecting trips to Hakone and Nikko.

In 1906, with the family's financial support, Handa travelled to England to study horticulture at Studley College in Warwickshire and it was while there that she was asked to help with designing the garden at Cowden. The contact probably came indirectly through Ella Du Cane, who had put Ella in touch with the American missionary Mary Denton, a teacher and fund raiser for Doshisha Women's College with extensive contacts among wealthy foreign visitors to Kyoto. Denton had got to know Handa when she was a student and suggested her to Ella, who wrote to her directly.[93]

The ground at Cowden earmarked by Ella as a possible site for the garden was a seven-acre marshy hollow that flooded in winter and was thickly overgrown with rushes. 'It had once been cut by a straight and strictly utilitarian ditch that was fed by a not-far-distant spring. This was an important item for the future safety of my garden, for in the land of the north with its impetuous and headstrong streams, rising almost without warning, dire disaster might otherwise have been its fate,' Ella wrote in her memoir. 'Realizing the possibilities of the somewhat uninteresting stretch of ground about the lake, and how, if the breath of life were breathed into it, it would become a living soul, I found a Japanese woman who could undertake the task and wave the magic wand.'[94]

Taki Handa's first visit to Cowden was in January 1908, when she spent a few days inspecting the site and discussing options with Ella, almost certainly with reference to Conder's *Landscape Gardening in Japan*. Although Handa was not a professional in the field, she had studied horticulture seriously and understood

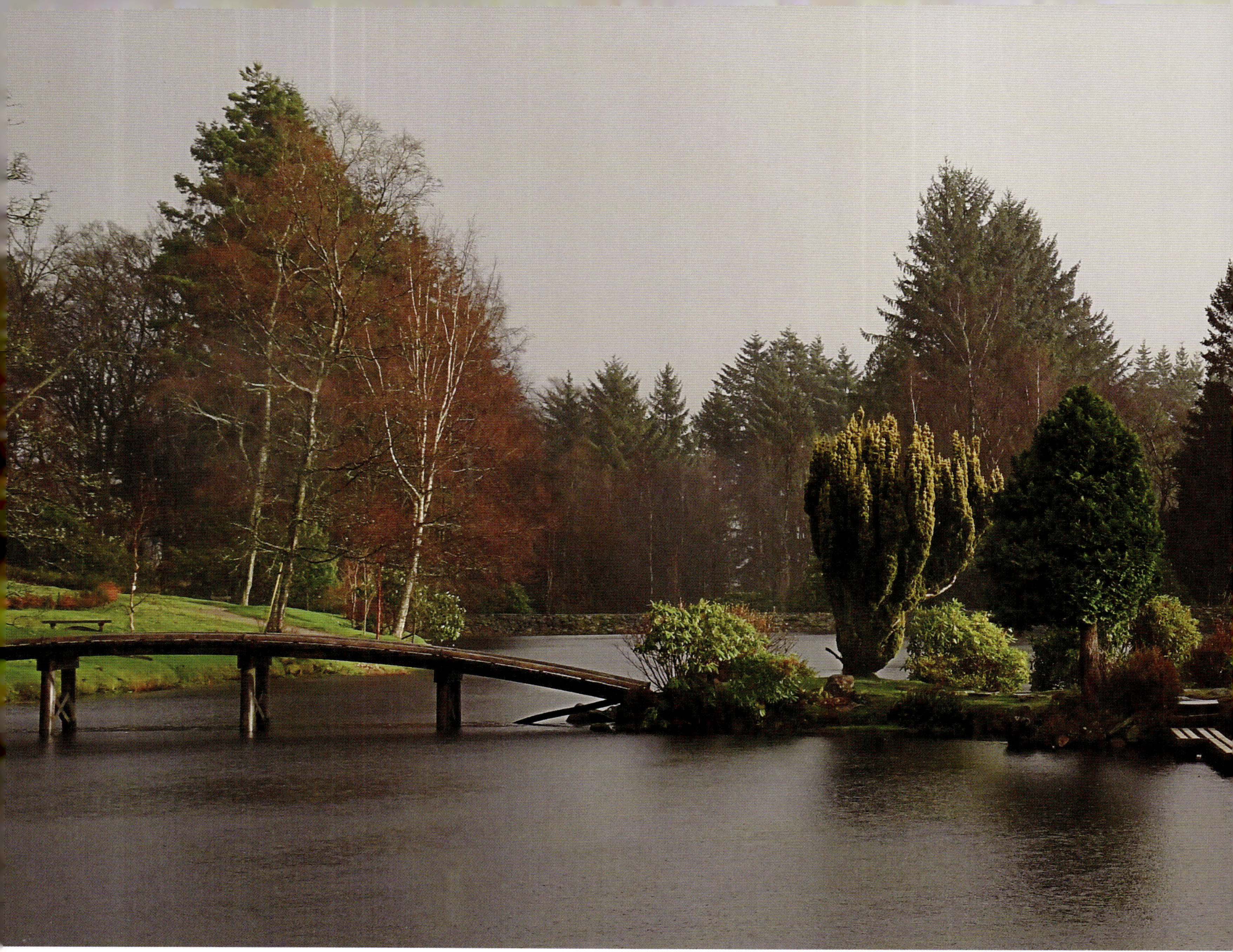

above The *azumaya* is a place from which to contemplate the garden, giving a clear view of the differently shaped bridges.

opposite right Taki Handa at Cowden Castle.

opposite far right A group of students at Studley College, Warwickshire. Taki Handa is at the front on the right.

following pages The Slopes of Fuji, the North Gate and the Master and Guest Isles in the garden at Cowden.

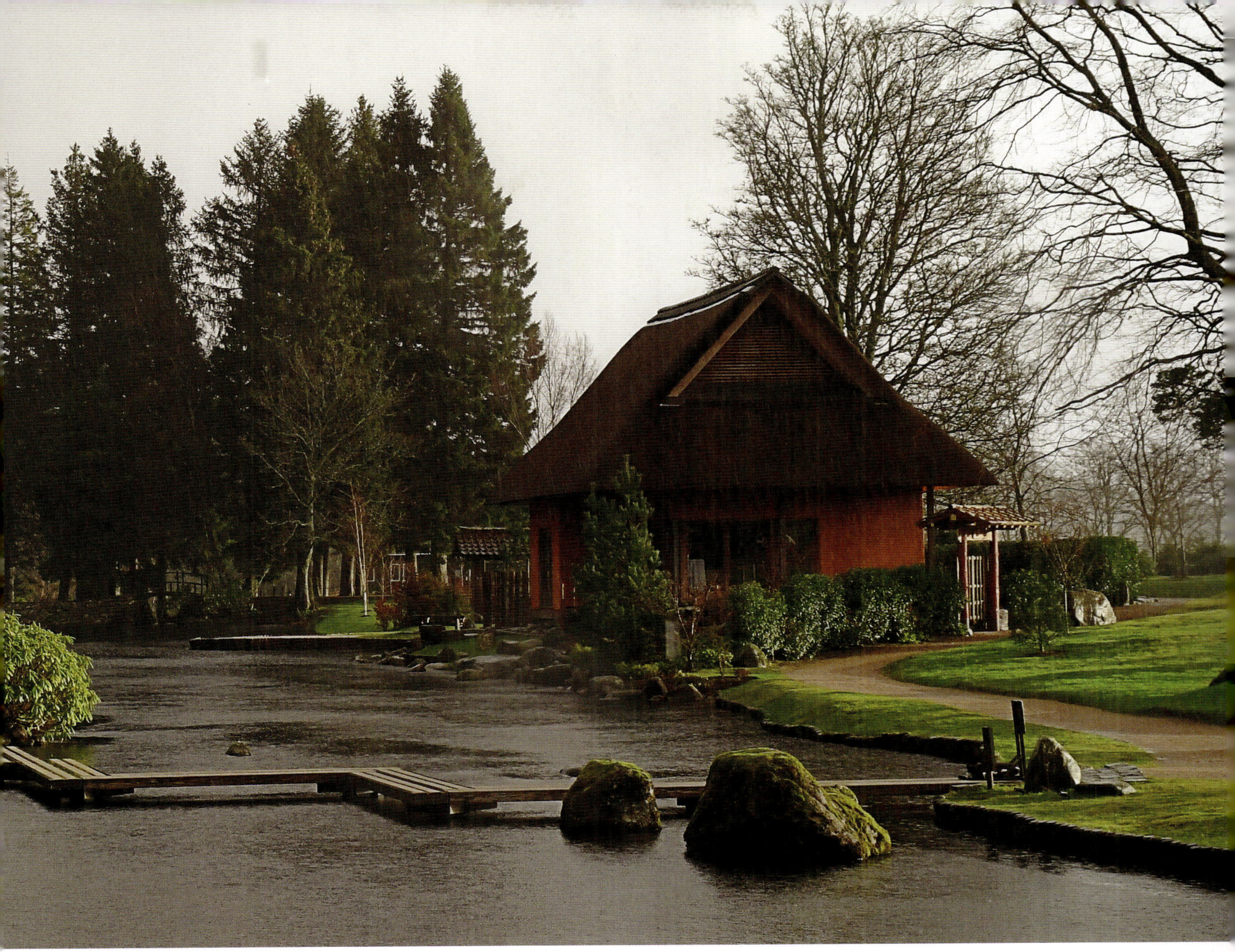

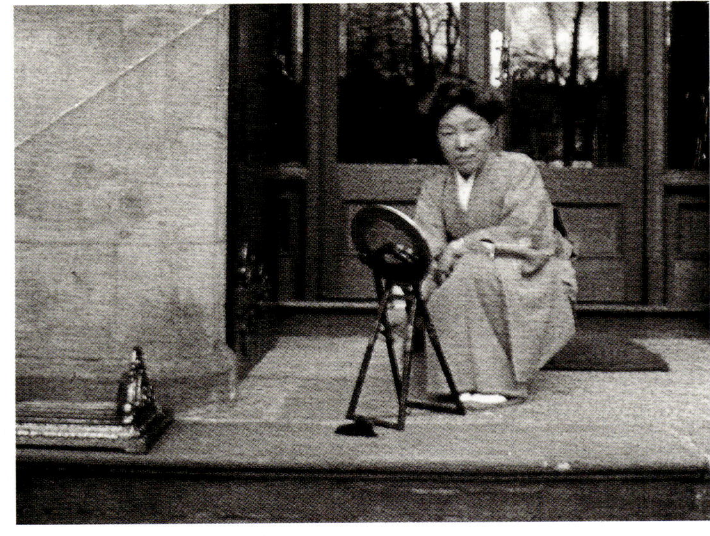
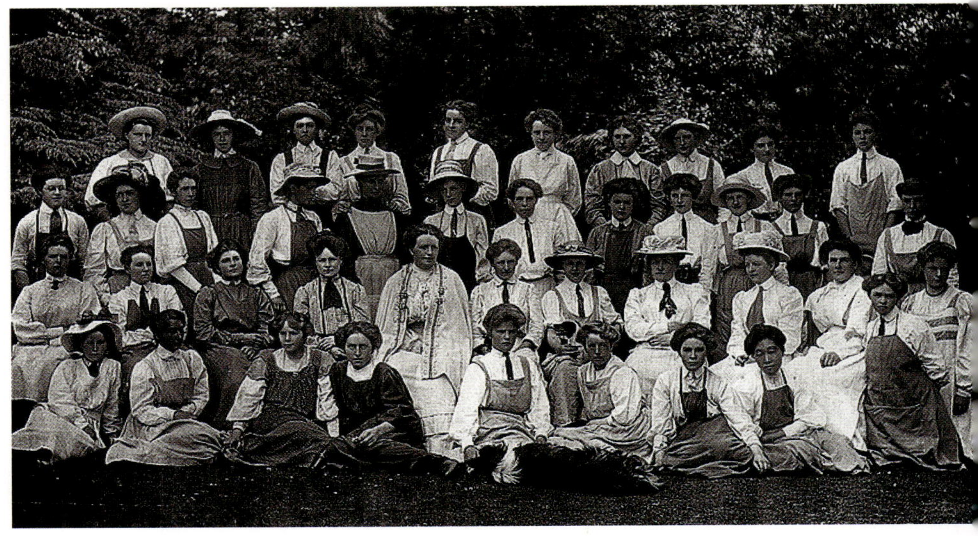

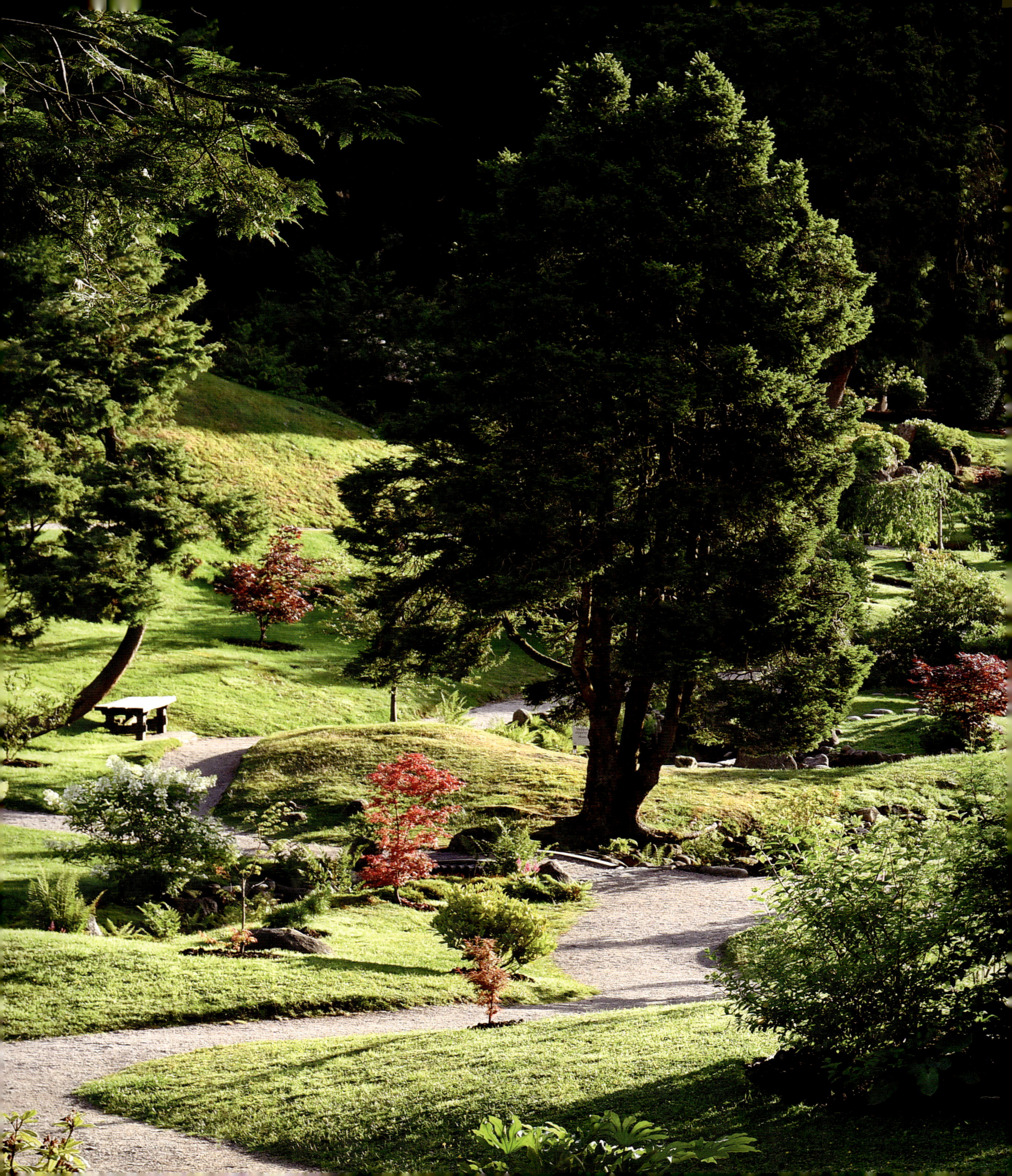

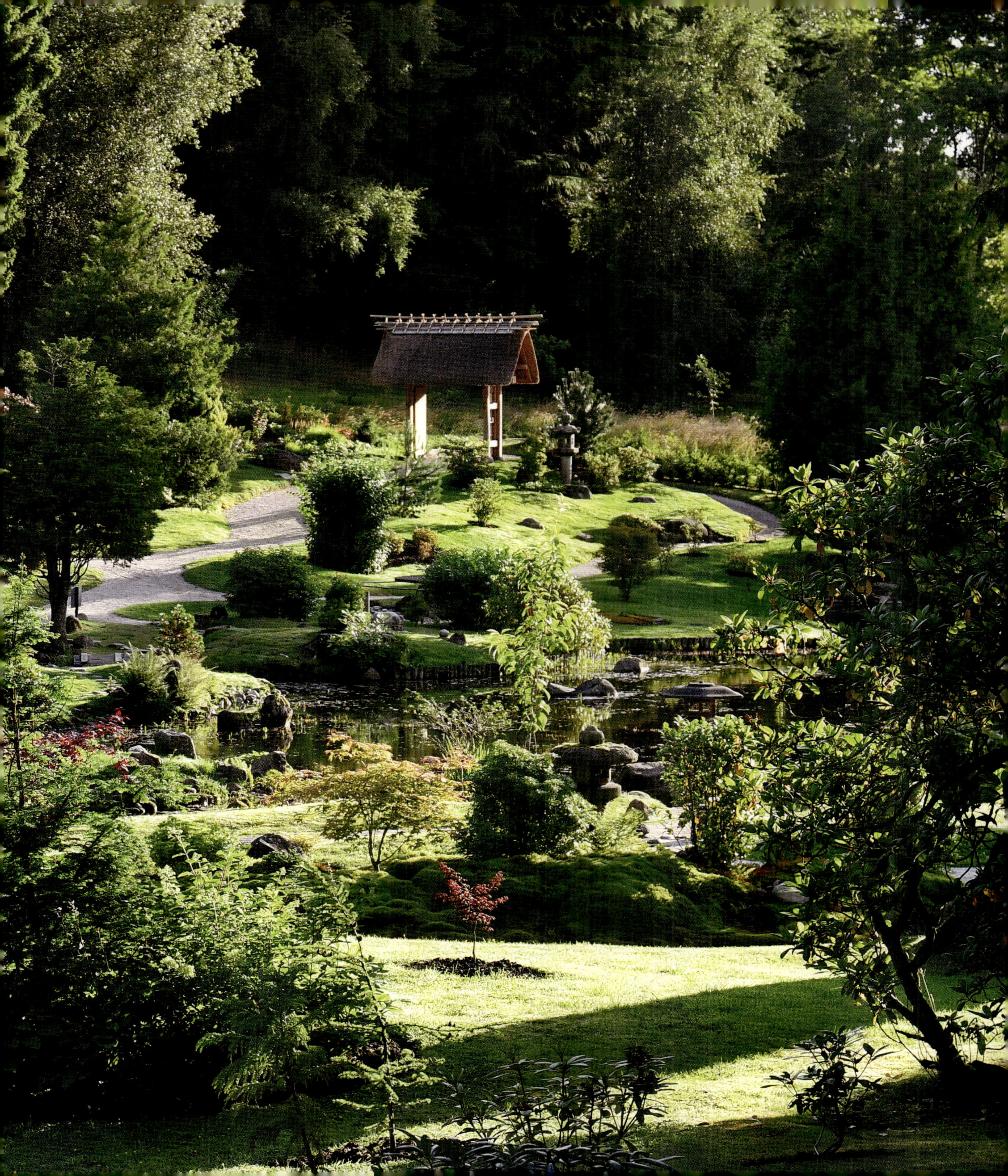

the nature of garden design; in addition to the writings of others, she also drew on her own instincts and feelings. Her work followed the principles of *shakkeishiki*, borrowing from the surrounding landscape and including a promenade around a lake – as seen in such notable examples as Koraku-En in Okayama, dating from the early 18th century, and the late 18th-century Kenroku-en in Kanazawa.

She returned North by train in April and instructed three estate workers to dam the ditch at the west of the garden to create a small lochan and to make an island with the extracted soil so that Ella could have her desired bridge. An island garden was developed and there was also a stroll garden and a teahouse garden with plants, shrubs, trees and traditional stone lanterns imported from Japan. Handa found some stones, mostly from a nearby old mill, and placed them around the site to make the setting more interesting and planted azaleas, box bushes, rhododendrons and lawned areas. 'The Scottish workers worked very hard for me and showed great interest in what I did and admired the way I changed the landscape,' she recorded in her autobiography, adding that she had to keep a close eye on what they were doing or things ended up looking more British than Japanese.[95]

Handa also described the Cowden design: 'It is divided into two styles of garden: hill garden (*tsuki niwa*) and flat garden (*hira niwa*). The central figure was a duck lake (*kamo yose no niwa*), which had a boat house as well. On the East side of the lake there is a natural slope made from the soil produced from digging out the lake. On the border of the slope and the flat land there is a small valley leading to the lake. There is a natural waterfall. I tried to make effective use of these natural features in my plan of a Japanese garden'.[96] The garden was called *Sharaku-En*, meaning a place of pleasure or delight.

After returning to Studley to finish her course, Handa made her final trip to Cowden in August 1908, during which she continued to develop her ideas. By this stage, the garden was about 80% complete, although the lanterns had not yet arrived from Japan so she carefully marked on the plan where they were to be situated. Ella meanwhile bought three stone lanterns for ¥185 (£18 6s) through the Tokyo antique dealer Daikokuya and arranged to have them shipped home. A note attached to the invoice reads: 'They are well covered with moss which will be dry after the long journey, kindly thoroughly wet them with warm water in which raw rice has been washed enough rice to make the water look milky. Frequent treatments with this rice water will keep them in good condition'.[97]

Handa returned to Japan that year and, in 1910, she married Seiichi Nakanome, a widowed physician with six children, with whom she would have two daughters. She taught at Doshisha Women's College in Kyoto until 1919, and then ran the family orchard at Mizusawa until 1932. She died in her mid-eighties in 1956.

After the garden at Cowden was completed in 1909, Ella maintained it herself with the help of her estate gardeners. However, at some point before 1925, she contacted 'Professor' Jiju Soya Suzuki for help in finding a suitable gardener. Suzuki had come

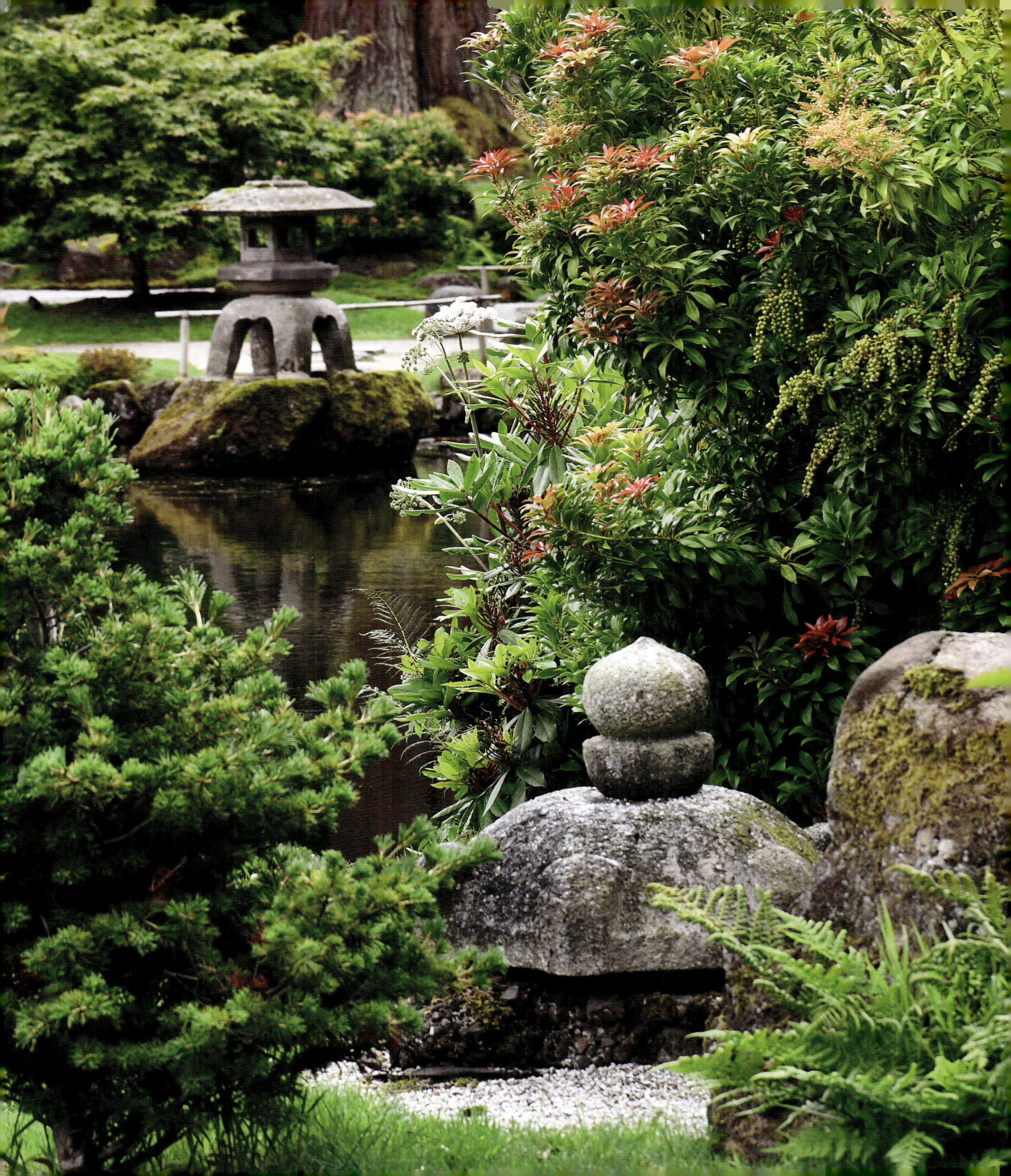

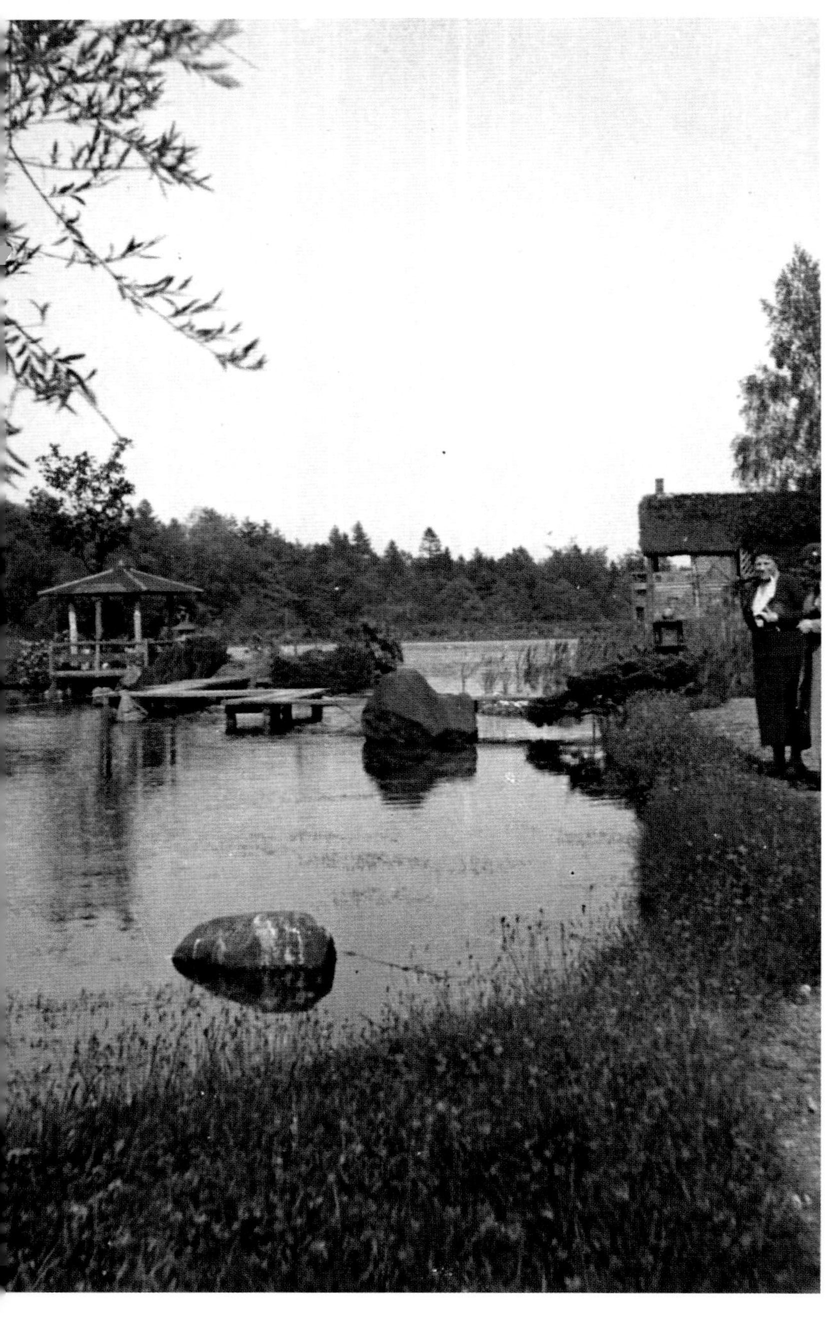

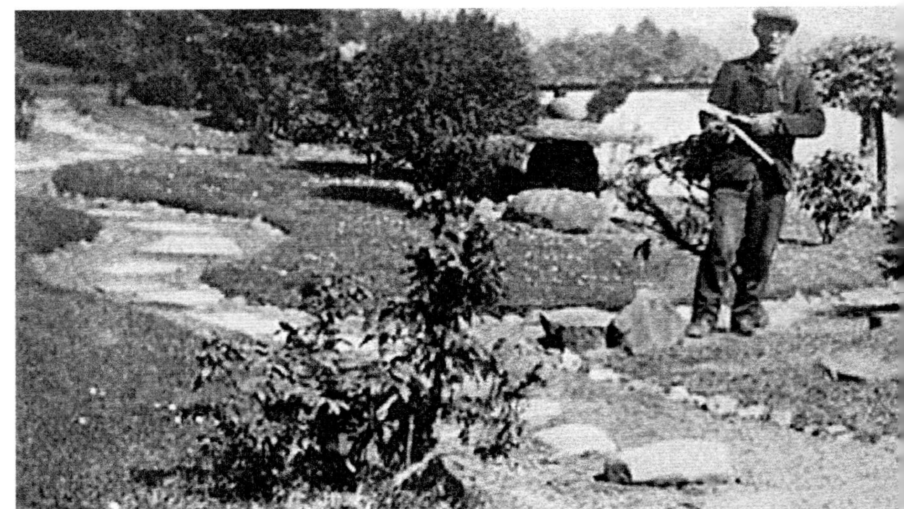

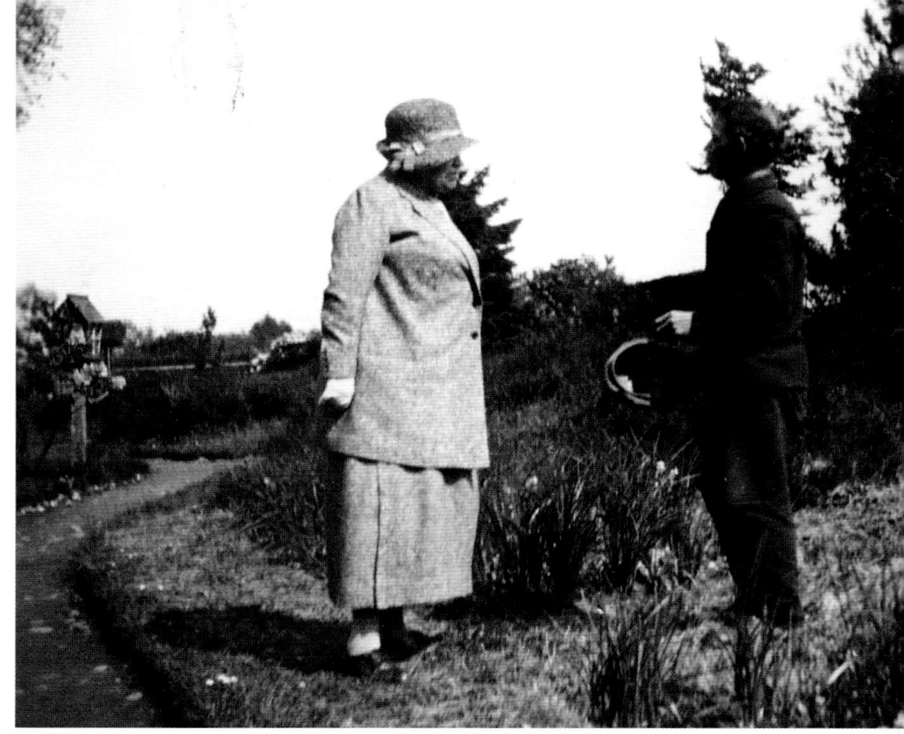

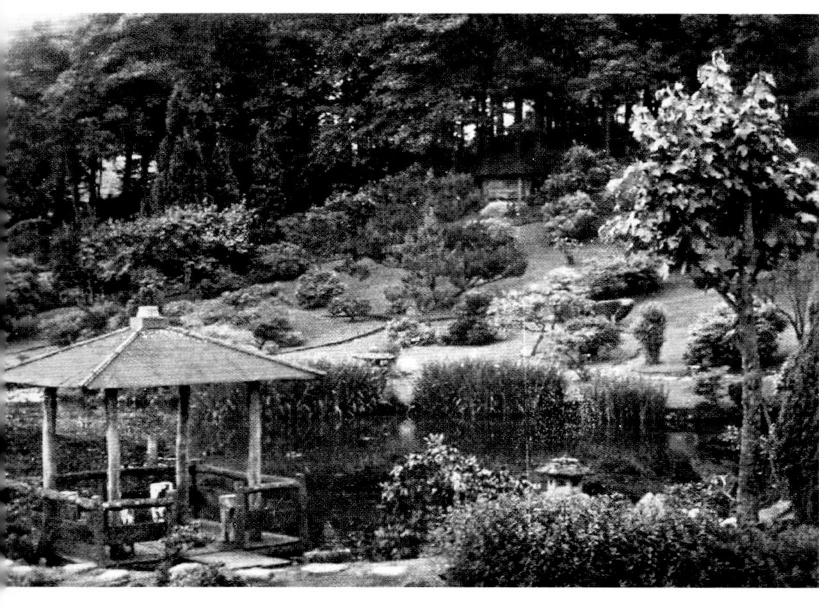

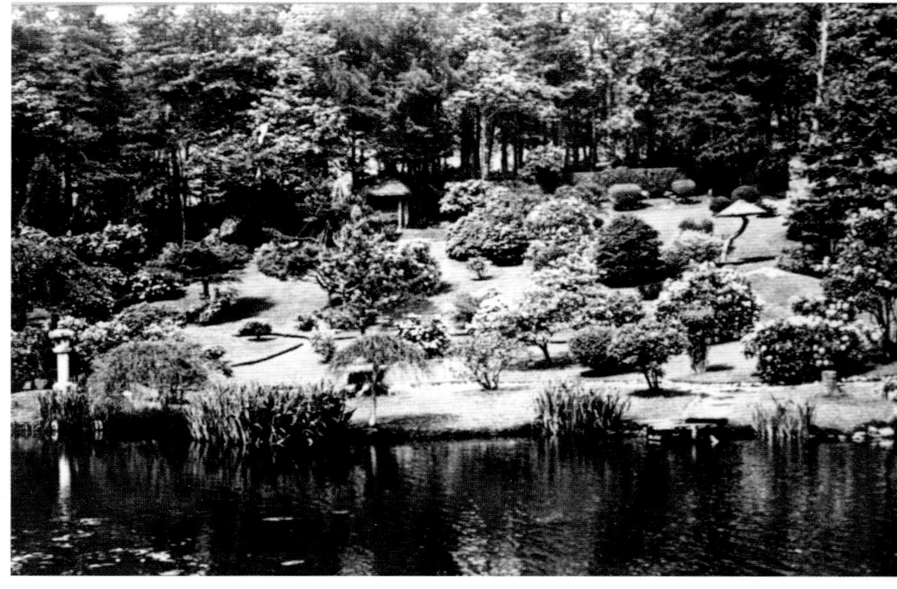

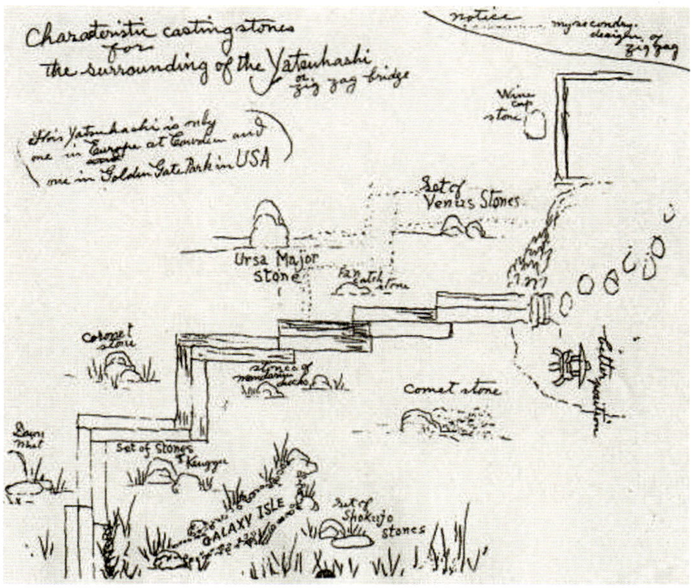

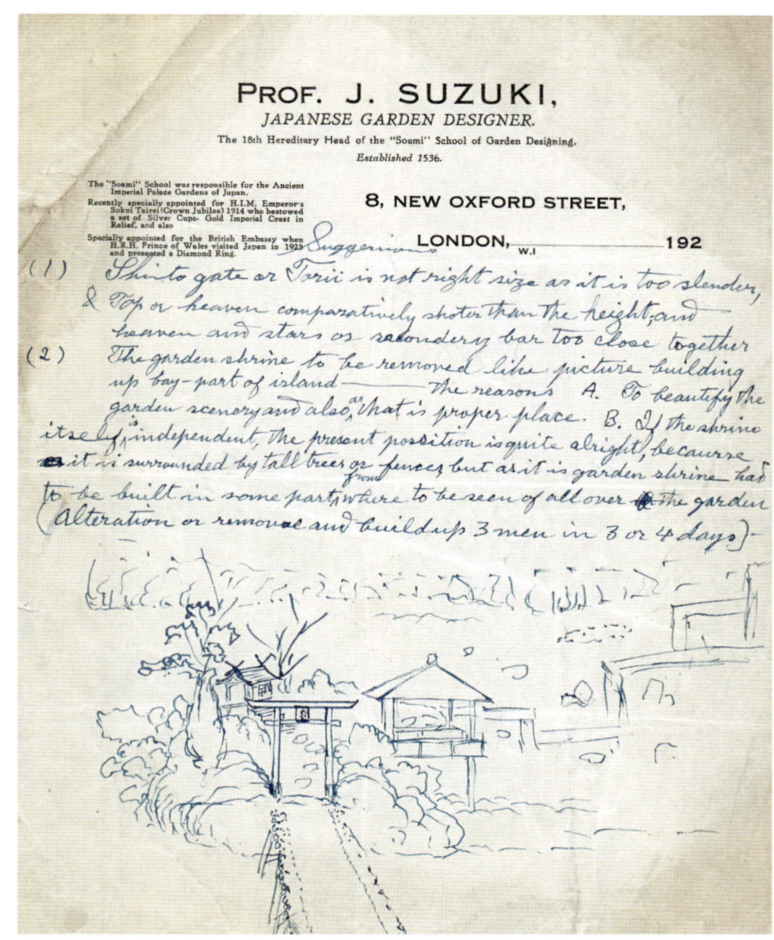

opposite, *clockwise from top left*
The newly installed *yatsuhashi*, seen here in 1934.
Shinzaburo Matsuo working in the garden.
Ella and Matsuo in conference.
The Slopes of Fuji in the 1930s.
The fishing hut in 1938.

this page, *clockwise from top left*
Professor Suzuki's original design for the *yatsuhashi* bridge.
Suzuki's suggestions for some alterations to beautify
the garden.
Suzuki's letter of recommendation suggesting to Ella that
she hire Matsuo to work in the garden.
Suzuki's design for the fishing hut.

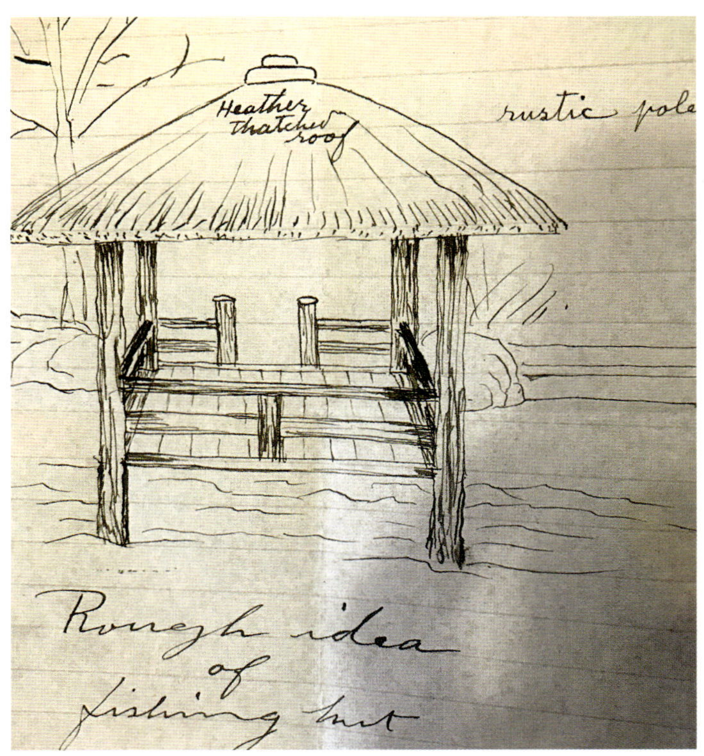

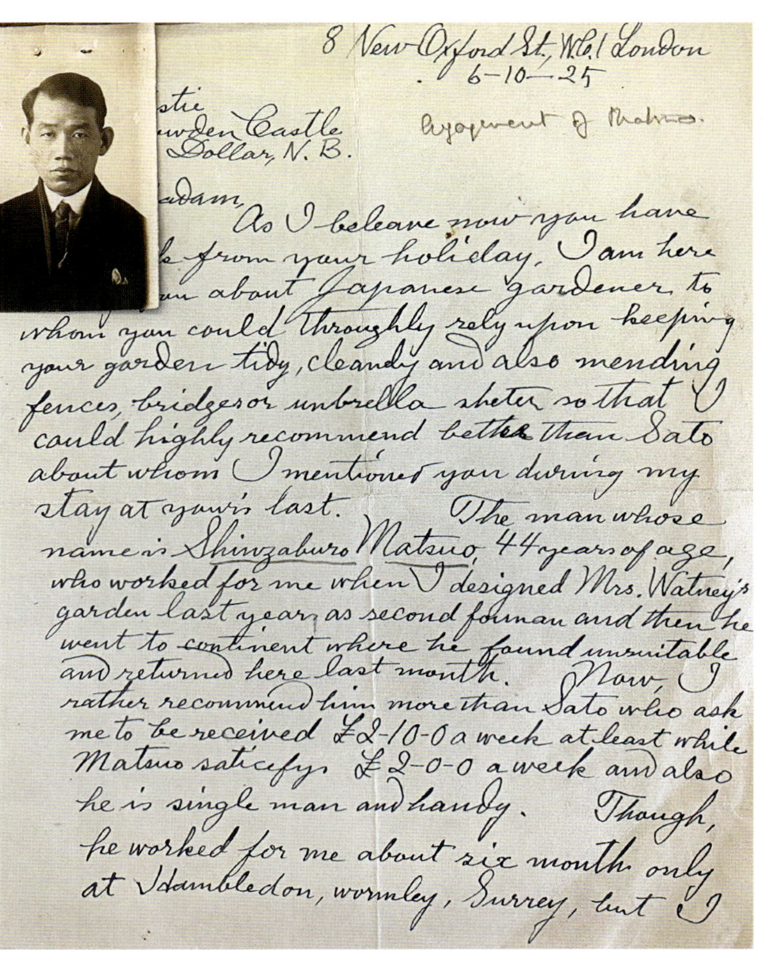

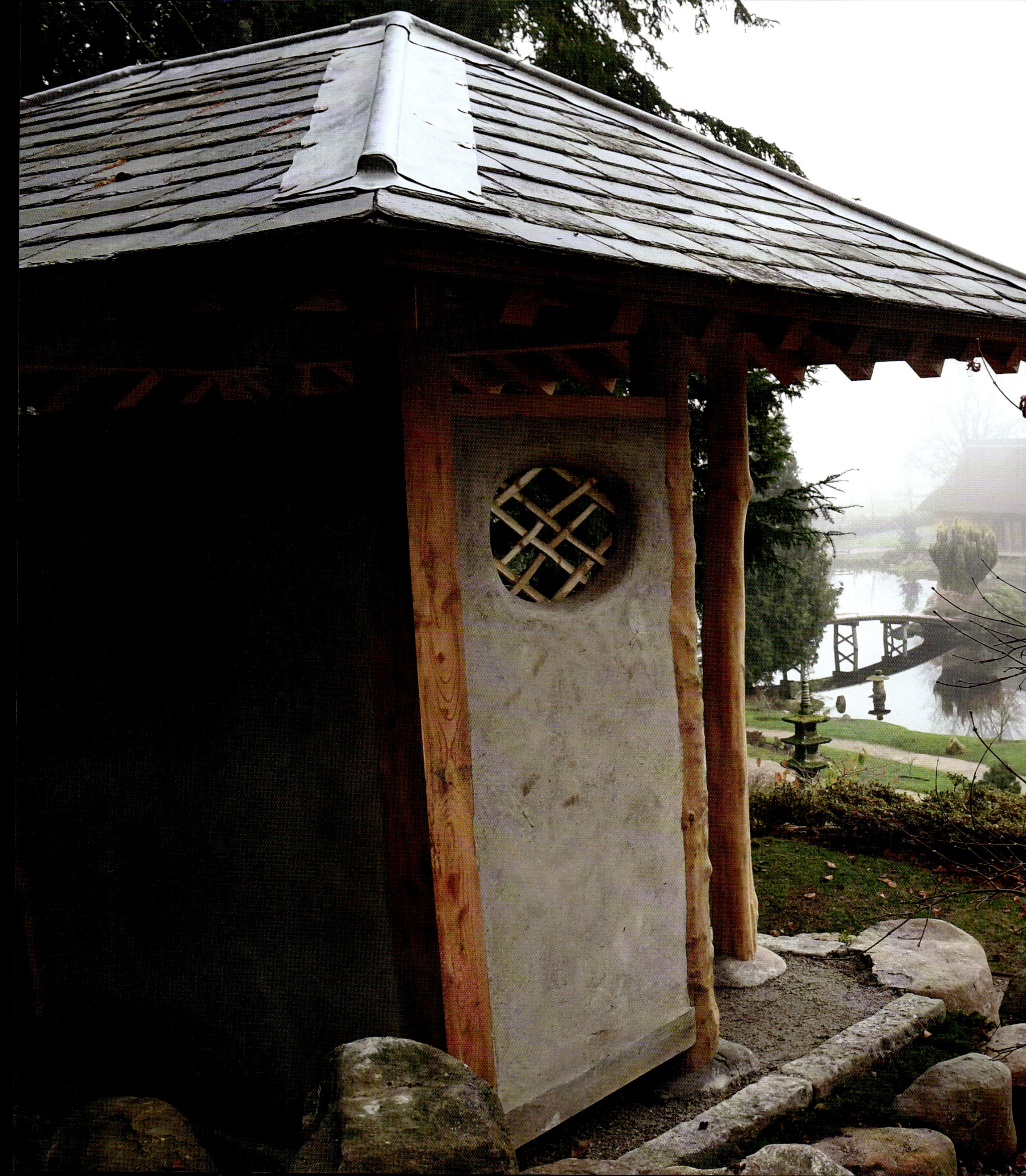

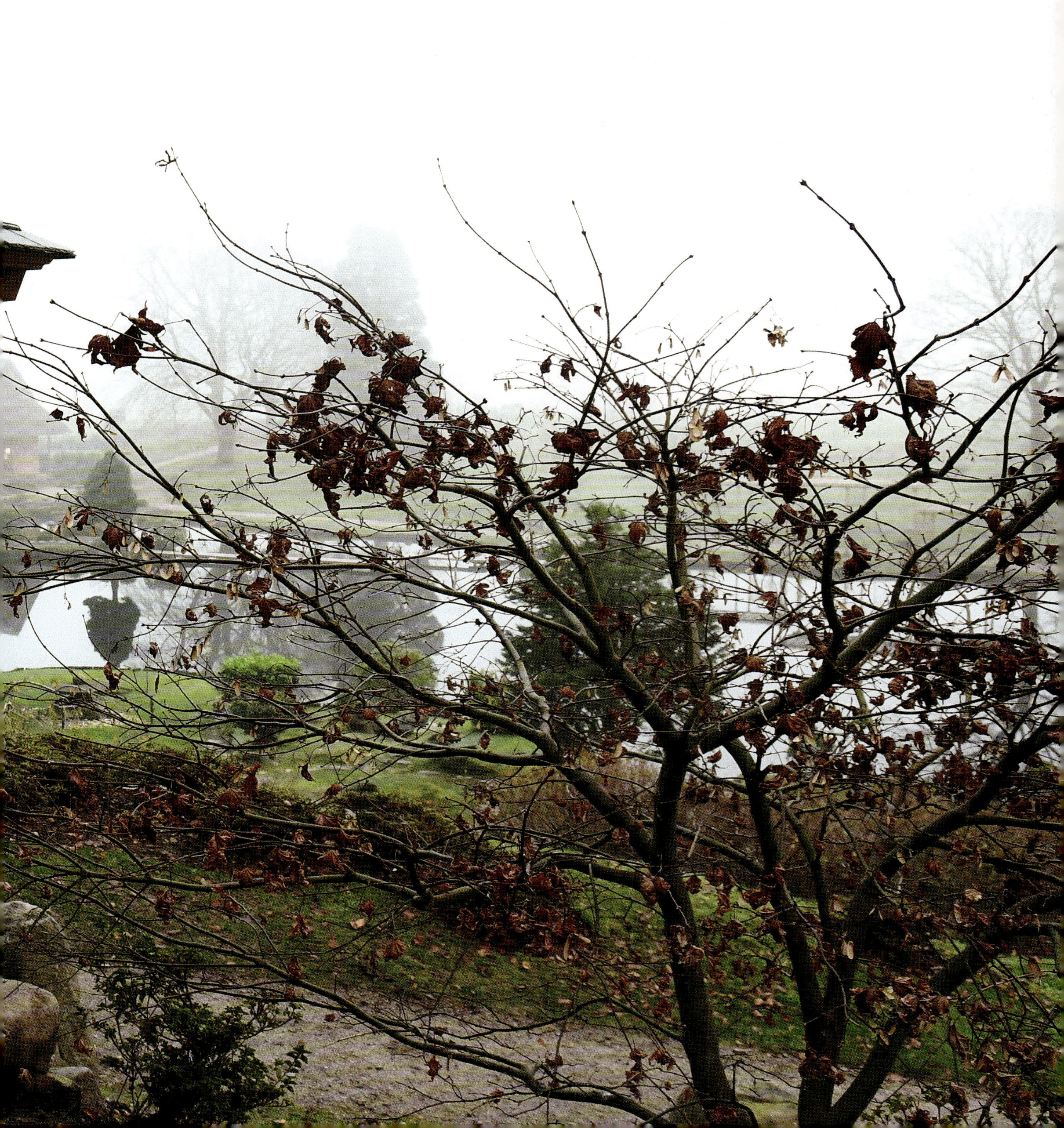

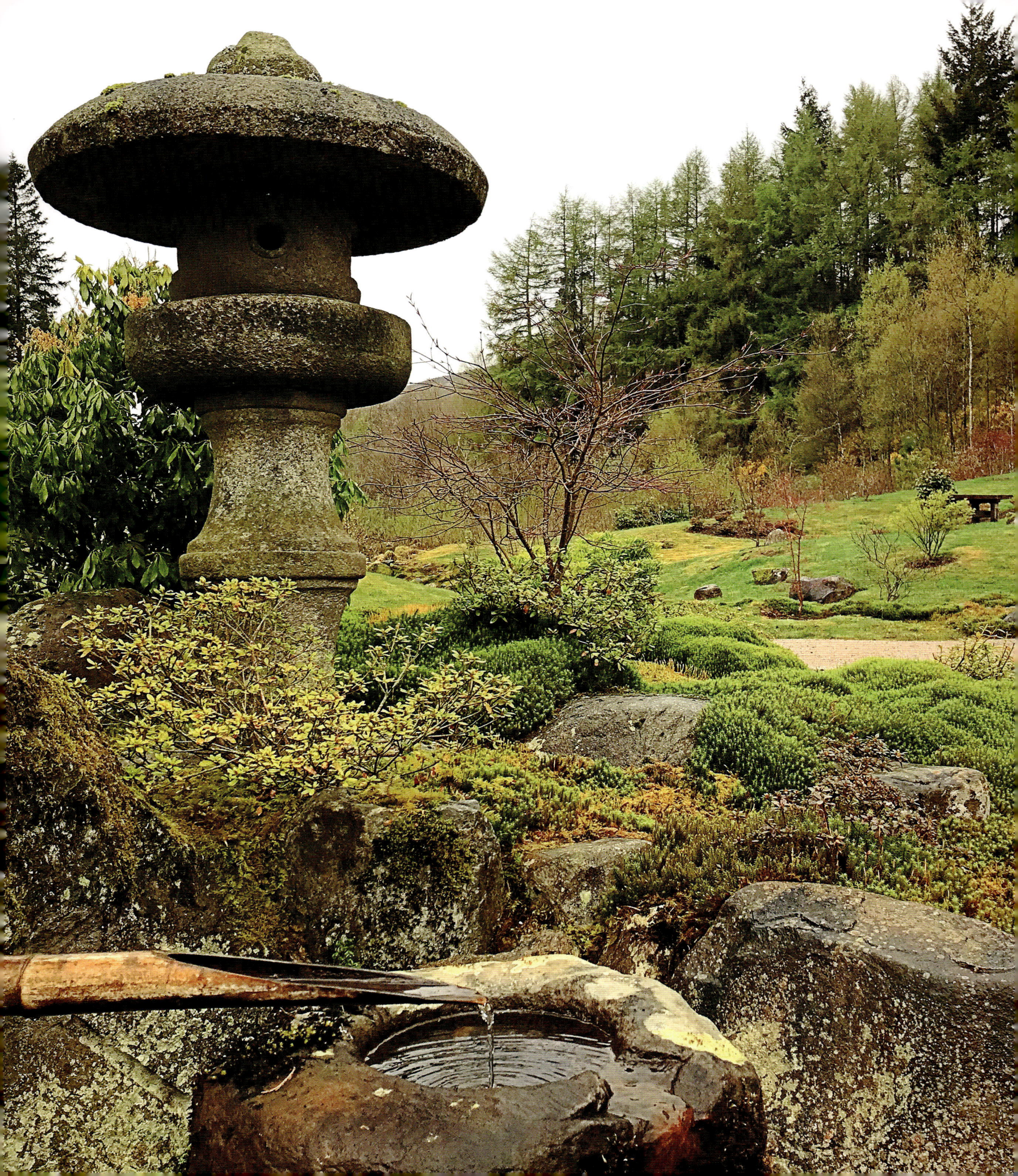

over from Japan in 1910 and set up in London as a Japanese garden consultant, describing himself as '18th Hereditary Head of the Soami School of Garden Designing Established 1536'. (His letterhead continued: 'The 'Soami' school was responsible for the Ancient Imperial Palace gardens and recently specially appointed for H.I.M Emperor's Sokui Tairei (Crown Jubilee) 1914 who bestowed a set of silver cups and also Specially appointed for British Embassy when H.R.H Prince of Wales visited Japan in 1923 and presented a diamond ring'.) It is not known when Suzuki came to Ella's attention, but by the time he wrote to her on October 6, 1925, he had clearly visited Cowden and conveyed his admiration for the garden to Handa, who was greatly reassured to receive this positive reaction from a professional, however exaggerated his qualifications.

The Japanese gardener whom Suzuki recommended to Ella was Shinzaburo Matsuo (1880–1937). He had come to England soon after the Great Kanto Earthquake of 1923, which had claimed his own family among its estimated 140,000 victims. At the time, there were only about 500 Japanese living in England, many of them in London, and it was probably through this small community that the paths of Matsuo and Suzuki crossed.

Suzuki assured Ella that Matsuo 'satisfys £2–0-0 a week and also he is single man and handy. Though he worked for me about six months only at Hambledon, Wormley, Surrey, but I found him faithful in the discharge of his duties, courteous and obliging in a taciturnity. I can assure you that if you engage him

he shall endeavour fully to justify it'.[98] Ella took him on and Matsuo worked at Cowden from 1925 until his death in 1937.

Meanwhile, her friendship with Professor Suzuki continued. He would send Ella letters (now in the National Library of Scotland), writing in his idiosyncratic style to suggest improvements. Notable among the additions Suzuki made to the garden over the years was a small shrine to Inari Ōkami, the rice god with attendant fox, and a fishing hut on the island in the middle of the lake (now lost); he also made alterations to the entrance gate.

In mid-December 1933, Suzuki wrote: 'my learned friend who visited the garden declares it is the best in the Western world, except for the bridge, which he thinks is just like flaw in precious gem. It is matter of regret to me to leave even one thing wrong in garden that I concerned [*sic*], and the fee is secondary importance. I leave it to you to consider alteration of wrong to right'.[99] The correspondence continues with drawings of his proposed replacement and more flattery: 'My Dear Miss Christie, as I know your appreciation of artistic gardens, your own among which is best in Europe and America by just little alteration that is I am sure you will be pleased to see a poetic scenery with the result and that it will be well worth the extra expense'.[100] The work to replace the southern arched bridge with a *yatsuhashi* bridge was complete by January 1933, to a cost of £22–10s. The invoice contained further emollient words to soften the financial blow: '8 plank bridge (only one in Europe or at yours) are belonged to the 'Hihō' or profound process of

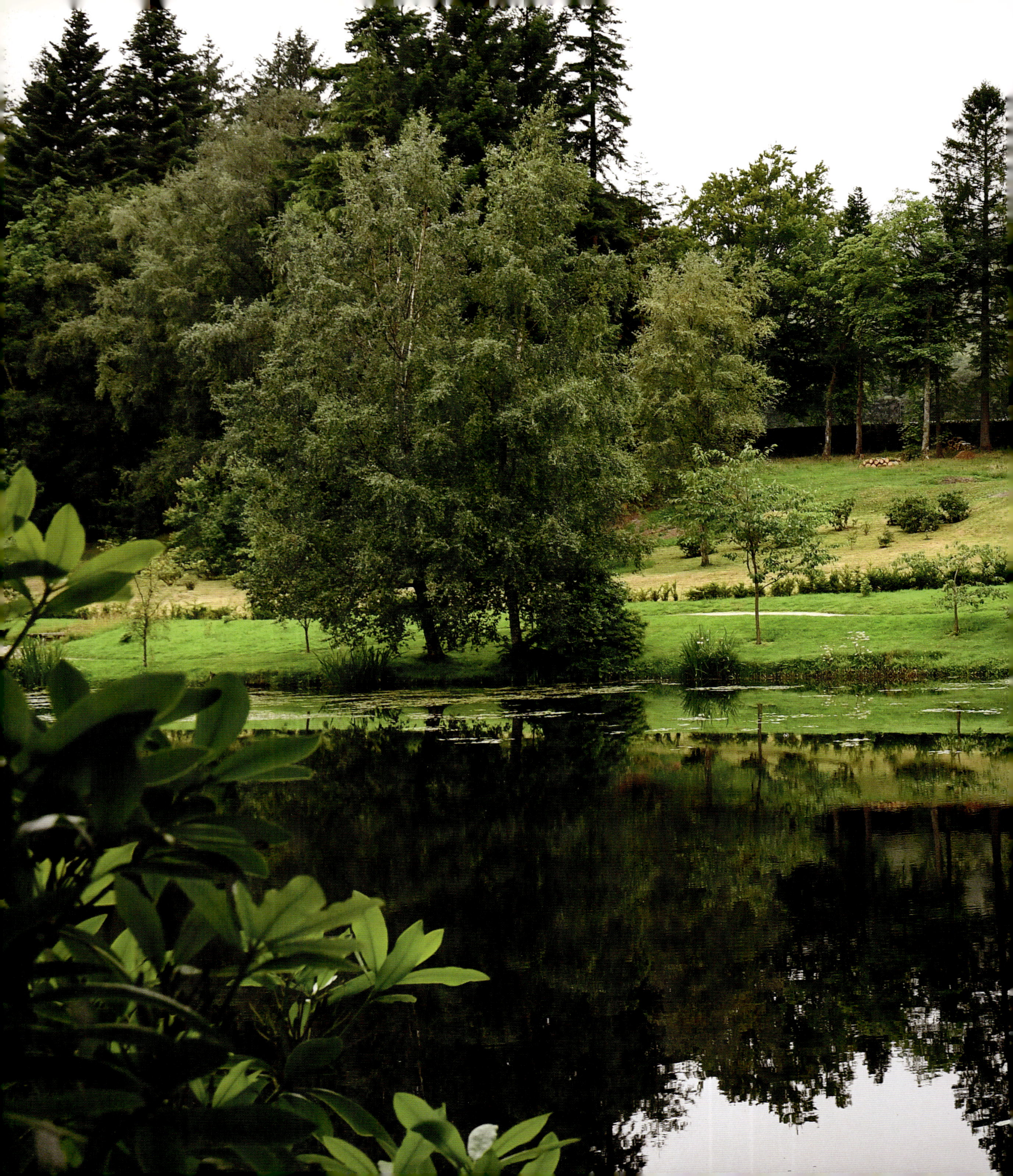

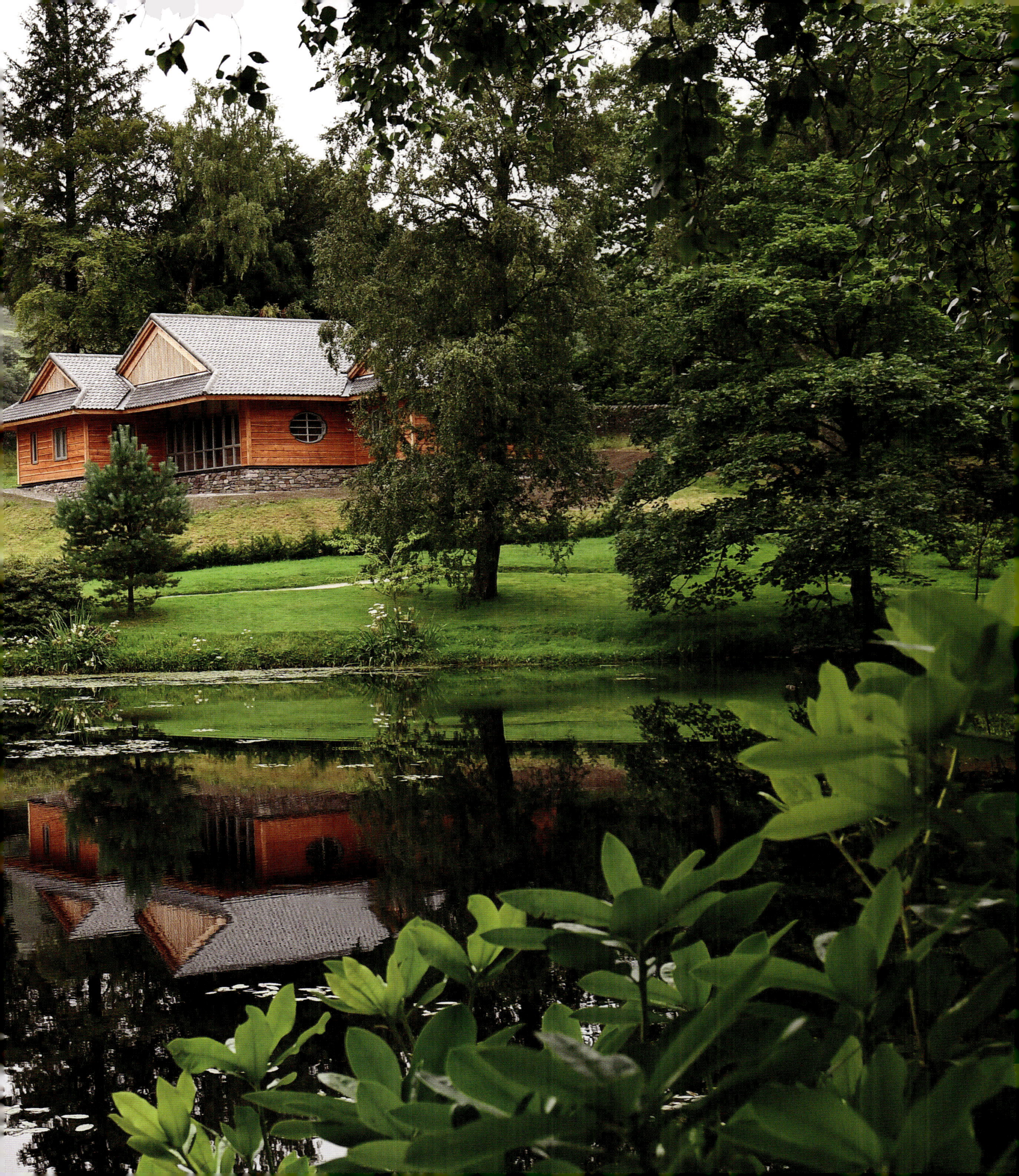

Japanese garden designing and we designer never lay out for ordinary client unless the high renumeration to be paid. I did it for you no matter the special reward'.[101]

Professor Suzuki had two families – one in Japan, which he visited infrequently owing to the cost of passage, and one with his British housekeeper in London. By 1933, he had returned to live in Japan, but he wrote to Ella from Nagoya in November complaining that there was no work and announcing his intention return to England. A letter thanking her for some magazines she'd sent to him reveals that, by February 1935, he was in Colindale hospital, Hendon, apparently recovering from a chill caught on the train returning from Cowden. In fact, he was suffering from TB. During his long convalescence, he asked Ella for some money in advance of work he might do for her, so that he could go somewhere warm to recuperate; Ella sent him five shillings. Sadly, Suzuki never fully recovered and he died in London on February 10, 1937. He was buried in Hendon Cemetery.

Back at Cowden, Matsuo lived in a two-room cottage just north of the garden and earned £2 a week. Margaret Borland from Dollar, daughter of Ella's chauffeur, described how he kept his house spotlessly clean and replaced all the internal doors with curtains to give it a more Japanese feel. Everything was of the best quality, the furniture, the beautiful linens, the food – not just rice, but a range of things; he was also very partial to a glass of fine wine. A lady recalled him coming to get some eggs one day; when she came back into the room with them, he'd disappeared.

Then she spotted him lying on his back inspecting the underside of her round pedestal table. Another lady found him doing much the same as he tried to work out how her husband's roll top desk worked. He was fascinated by furniture. His own pieces, including three chiming Westminster clocks, were beautiful, but in his last year, he began to sell them off, worried that he would not be able to keep them clean.

When Suzuki described Matsuo as 'obliging in his taciturnity,' he was probably referring to the gardener's patchy English. Ella had more than enough experience of being in places where she couldn't speak the language and was a master at communicating things that needed to be conveyed. 'By a process of what someone well called "sighs and wonders" they planned the work, and one watcher at least has a lasting memory of the two standing together by a bed of irises, deep in a kind of hybrid conversation; he smiling and small, with his hat in his hand (he always snatched it off when he saw her coming) she as usual a little larger than life in her homespun garden coat lined with the tummies of her own grey squirrels, and an amorphous felt that had known the suns of Samarkand as well as the snows of Scotland. Both were attended by a respectful half-circle of black ducks, their beady eyes hopefully fixed on the toy bucket of bread which Ella always carried.'[102]

Matsuo's dress could also be a little eccentric. He patched a hole in the back of a waterproof that Ella had given him with a contrasting fabric and cut off the bottom to make a frill to go round his neck. Once in a while, he would don his traditional Japanese

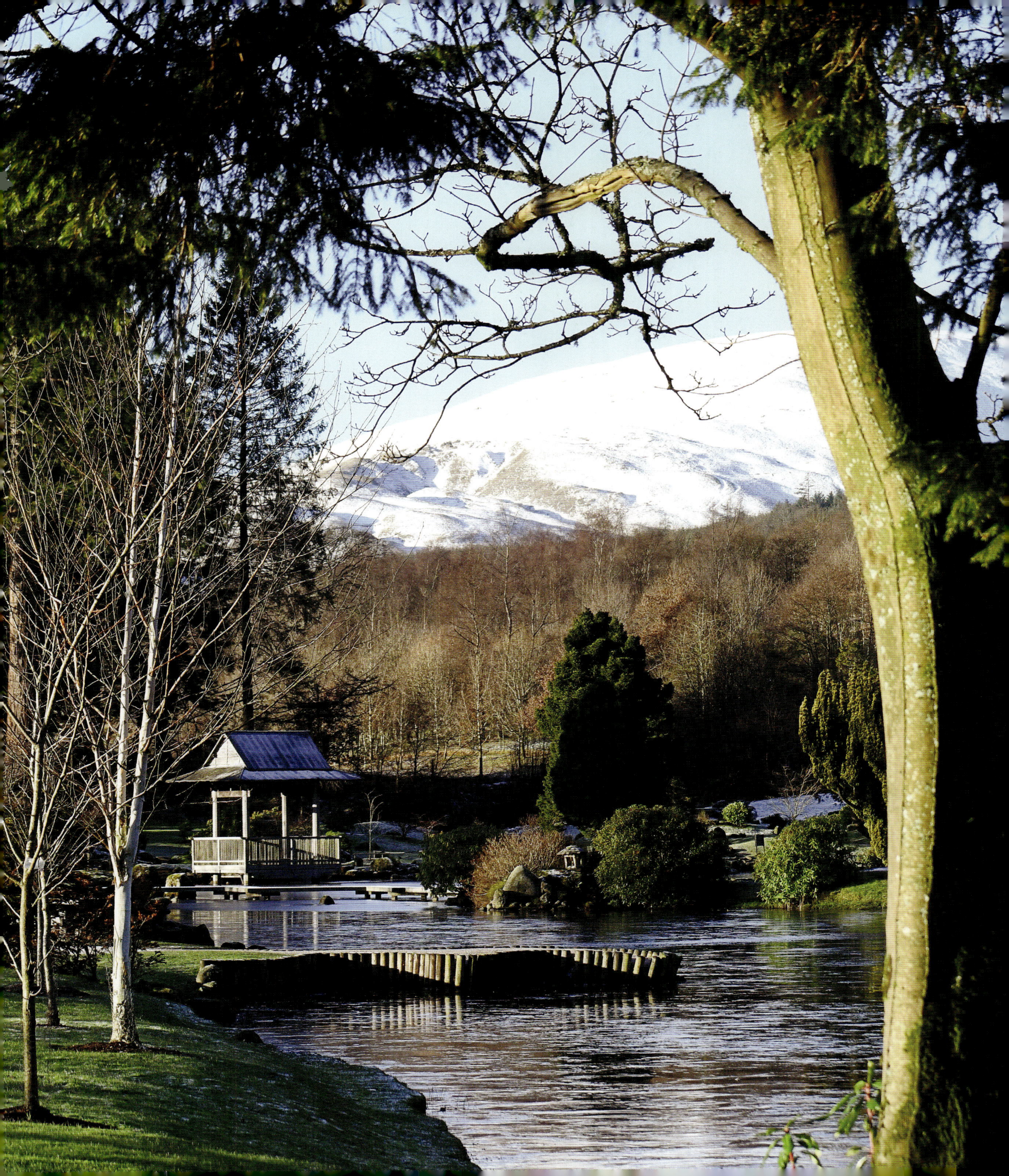

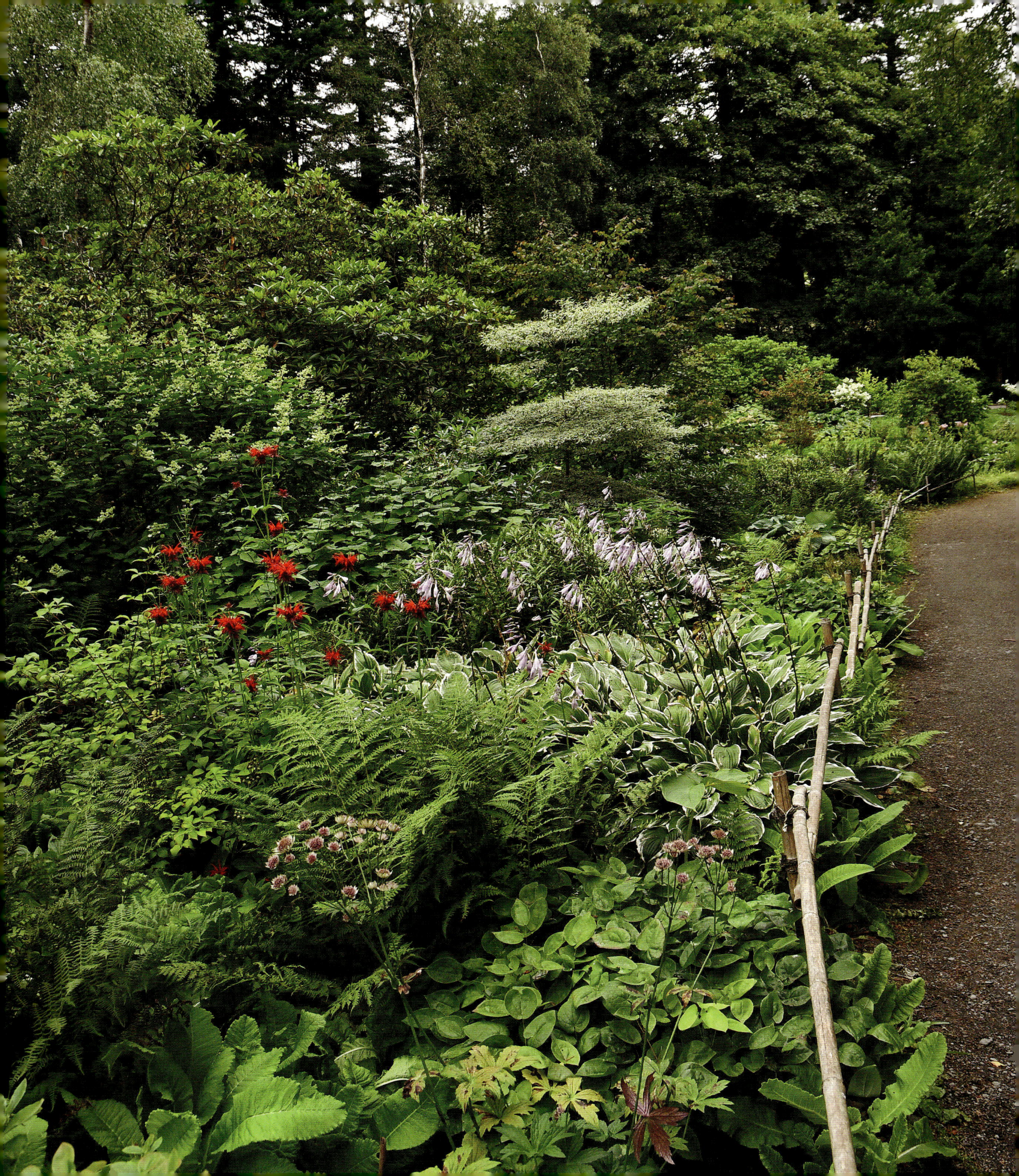

garments: wide, pleated trousers tucked into long socks, white gloves and an over *kimono* of bright colours, topped off with a velour hat. Given Ella's reputation for having a rich variety of friends, many were under the impression that he was the Emperor of Japan on an incognito visit to Cowden.

People remembered his kindness to children, how he delighted in giving them little gifts and was always singing. As a boy, Mr Borland was treated to a demonstration of Japanese dance with accompanying explanations. Matsuo was always at pains to explain the characters written on the boards he hung at various points in the garden. Inside the Inari shrine were pottery animals, a polished brass shield and a central earthenware bowl containing a scroll inscribed in Japanese. He explained how the bell that hung outside with a thick ribbon-pull should be rung as the doors of the shrine were opened and you bowed three times. If your face was reflected in the brass shield, then you were good; if not, you were bad.

Eventually, he decided to find a wife and went down to London to meet a Japanese lady to whom he had been introduced by Suzuki. Her name was Hiro Katsuda and their meeting at the YMCA in Tottenham Court Road went well; Matsuo returned to Scotland with the intention of asking her to marry him. Soon after arriving home, however, he developed a gastric disorder, possibly a perforated ulcer. Sadly, this was to prove fatal. Matsuo died on October 20, 1937, the same year as Suzuki, and was buried in the Muckhart churchyard. Had he lived a few years longer, into the Second World War, he would undoubtably have been interned along with other Japanese residents in Britain.

'I am of the opinion that Mr Matsuo was a very kind and generous person, sometimes comical but always likeable and after his death the Japanese garden was never quite the same,' Mrs Borland concluded.[103] Ella's hand-written tribute is among her papers in the National Library of Scotland:

'To the many who have visited the Japanese garden at Cowden a loss will be keenly felt in the knowledge that its faithful keeper is there no more. For nearly 15 years he cared for it and its interests with a fidelity that is rarely to be found and with the artistic skill so much a possession of his race. In his strange garb which he seldom varied even in the summer heat or winter cold unless in the latter season when he sported an aviator's leather helmet or in the former when summer was notified by a gay duster suspended from the back of an antiquated velour hat, his figure much resembled a little brown grub as he patiently extracted the minutest weeds and one scarcely realised the beauty of spirit that dwelled within that humble exterior. His courtesy of manner and his kindness and generosity of heart especially to children revealed this. He was of the Christian faith which he lived up to the highest ethics of his country's own standard. These were embodied in a paper written by himself in English now quoted in memory of one who remained faithful unto death and as an epitome of a perfect life.'

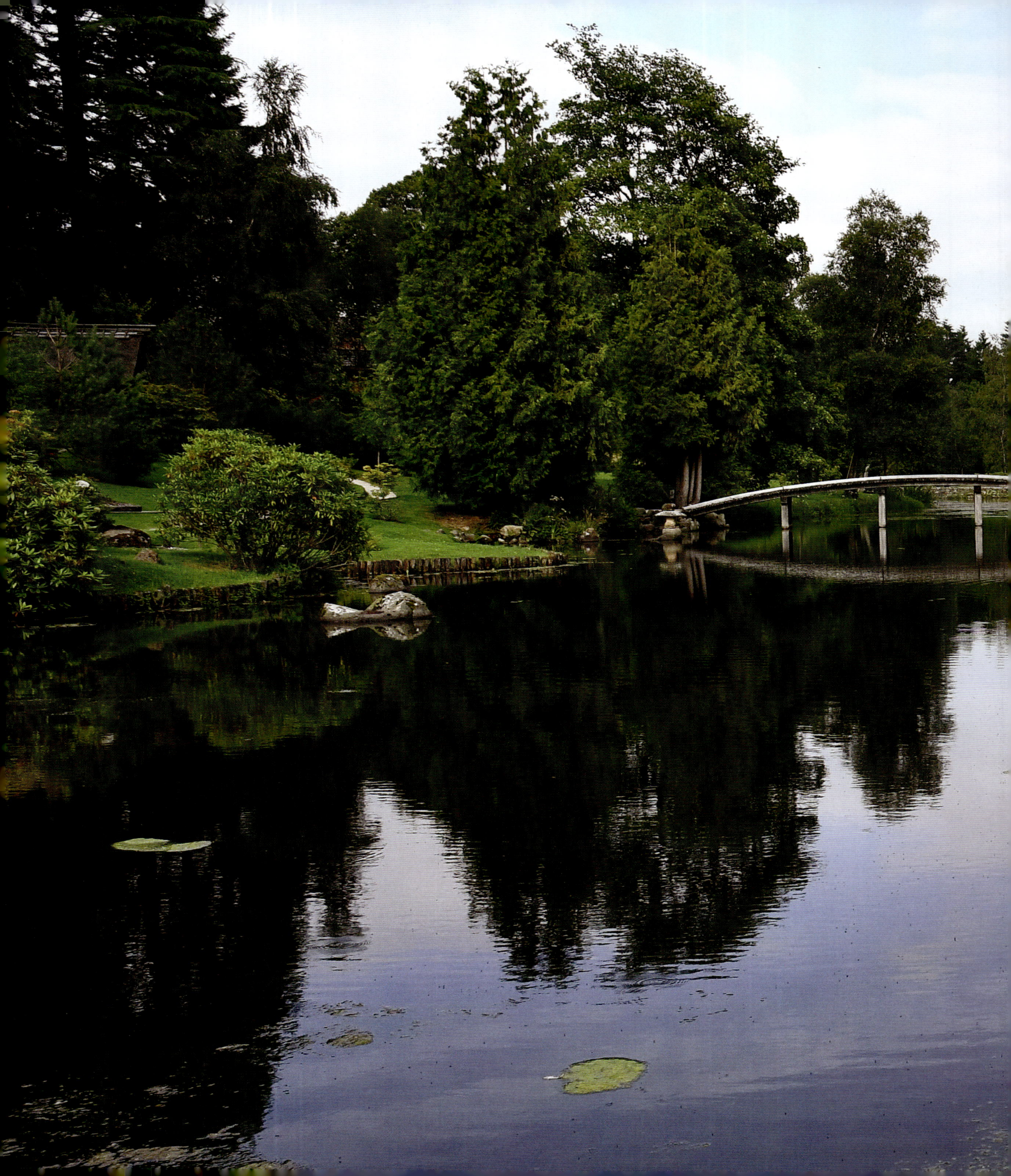

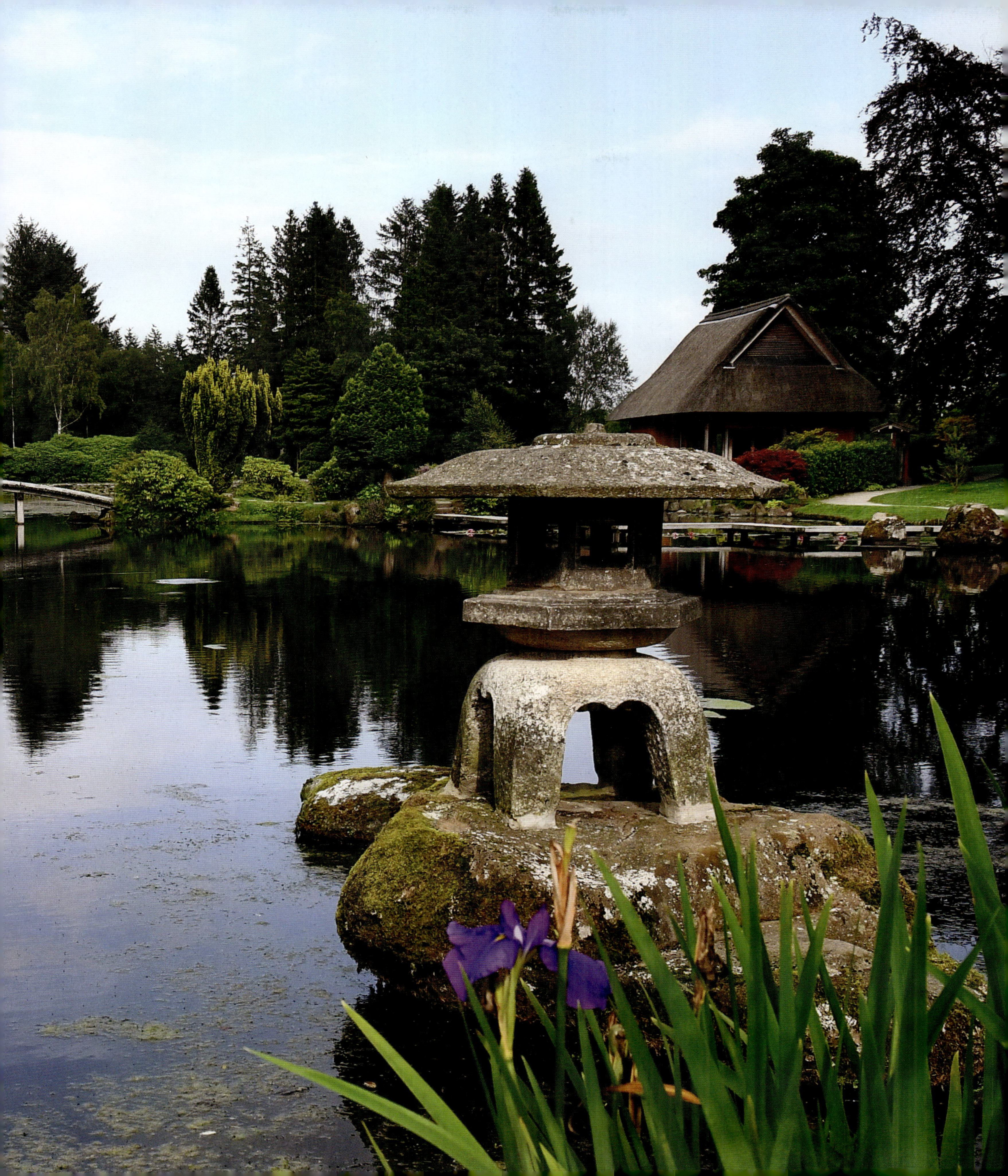

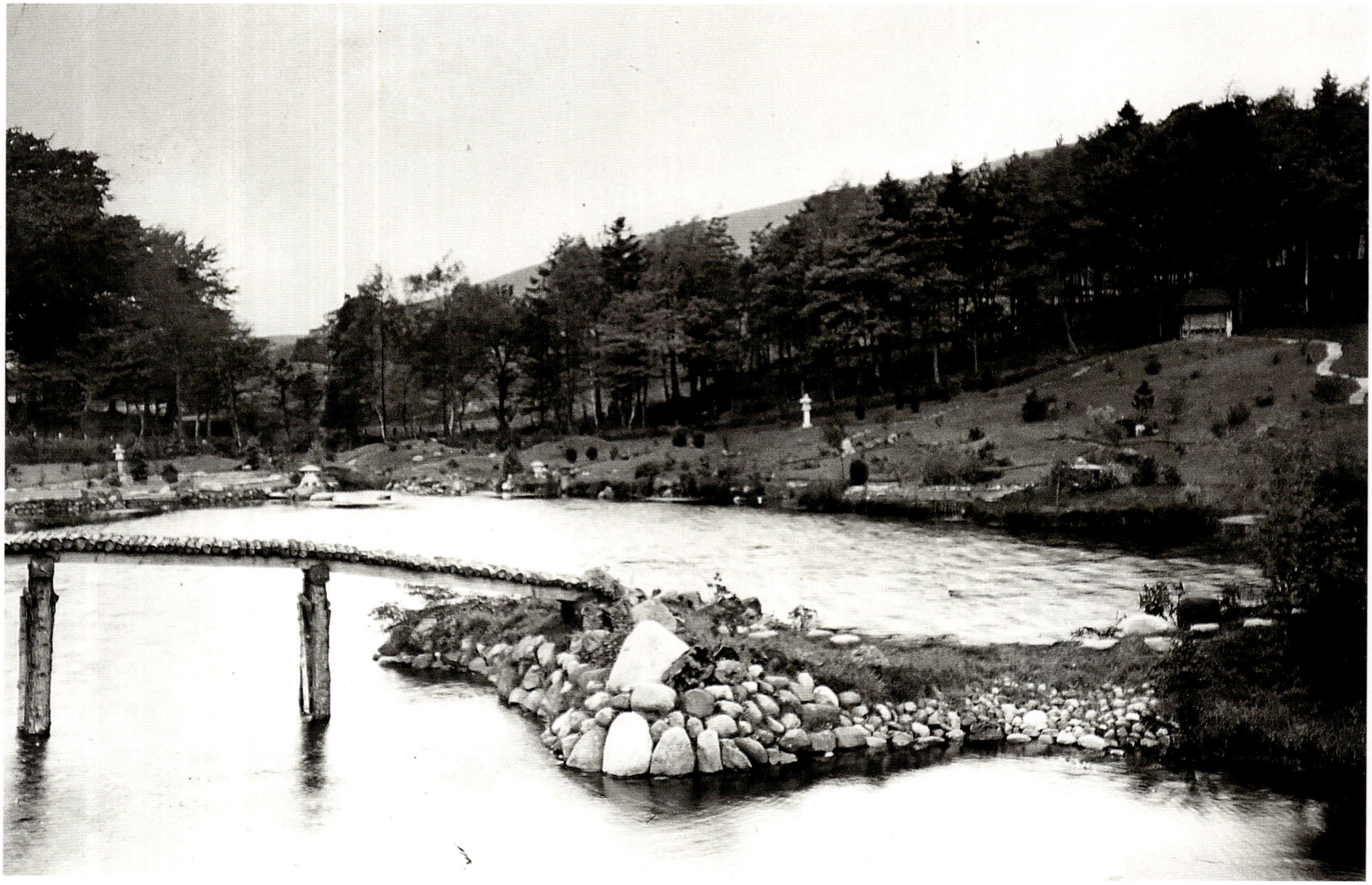

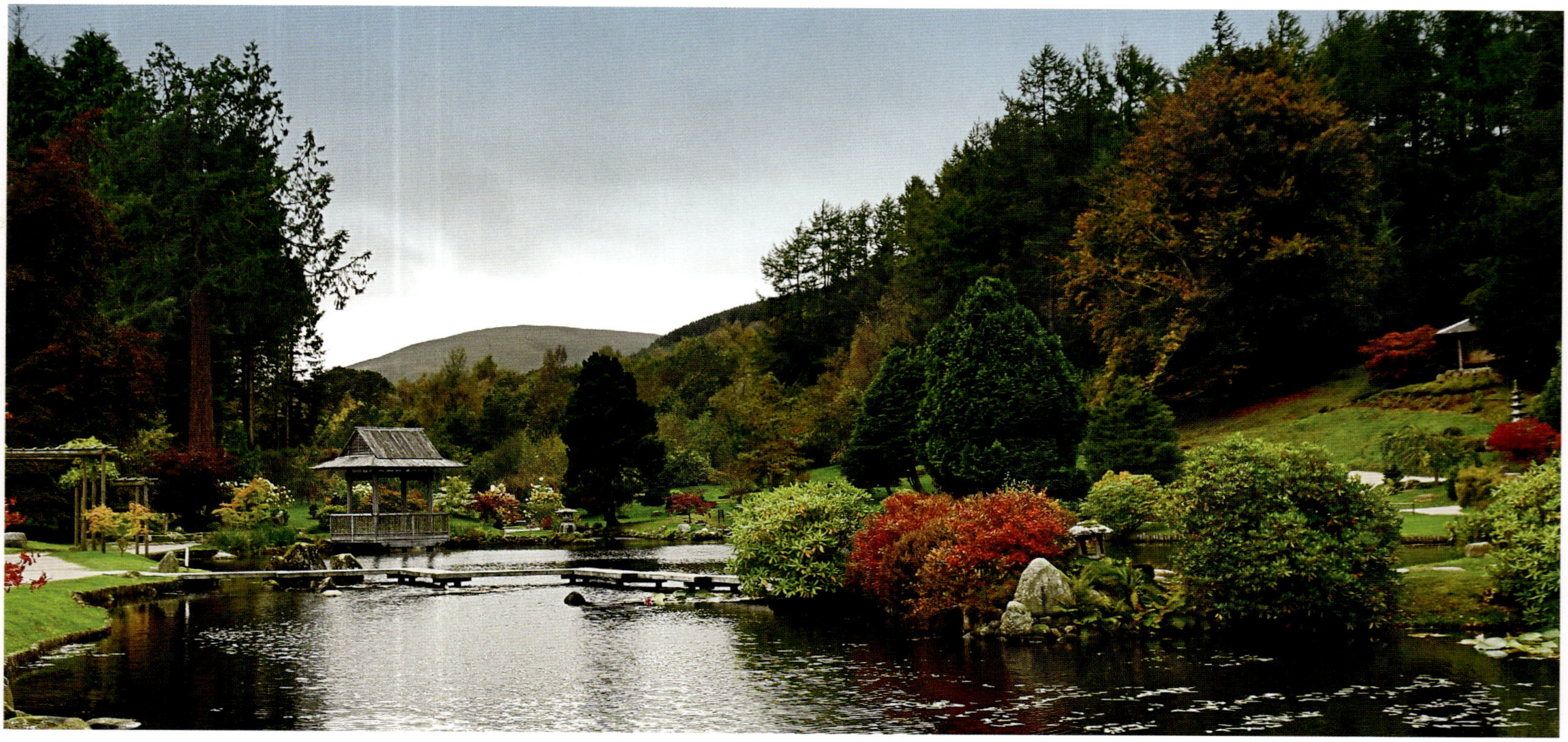

A Tour of the Garden and its Structures

left, above and below
The garden is designed to fit into the surrounding landscape and to borrow from it. These two photographs, taken from roughly the same position 100 years apart, illustrate how it sits beneath the Ochil Hills, with the trees – including the now giant redwoods – acting as a demarcation zone rather than a barrier. There are several gates, including the Welcome Gate on the site of the original entrance gate and the North Gate, or 'sweeping gate', a faithful copy of the original at the exit on the slope above the lake.

right, above and below
In the black-and-white photograph, the entrance gate is visible on the left, almost behind the tree. The recent photograph of much the same view shows the thatched Garden Pavilion that replaced the boathouse as the garden's main structure, its design based on the *shōkin-tei*, or teahouse, at the Katsura Imperial Villa in Kyoto.

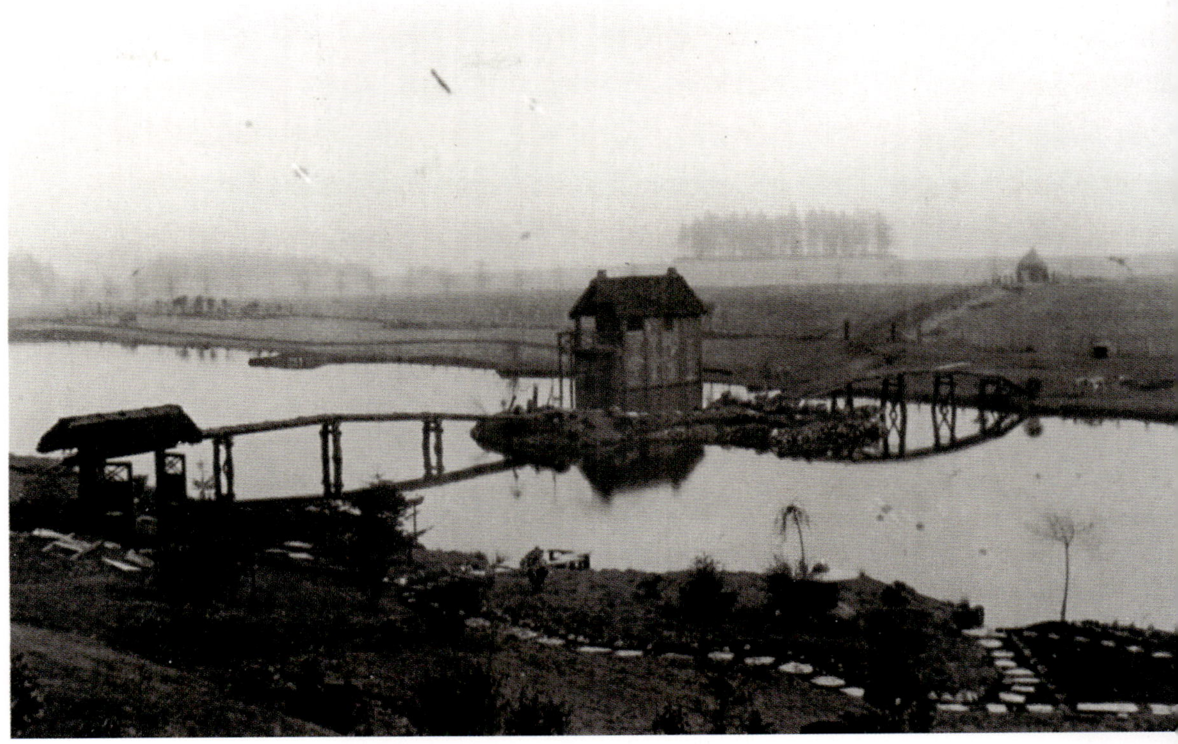

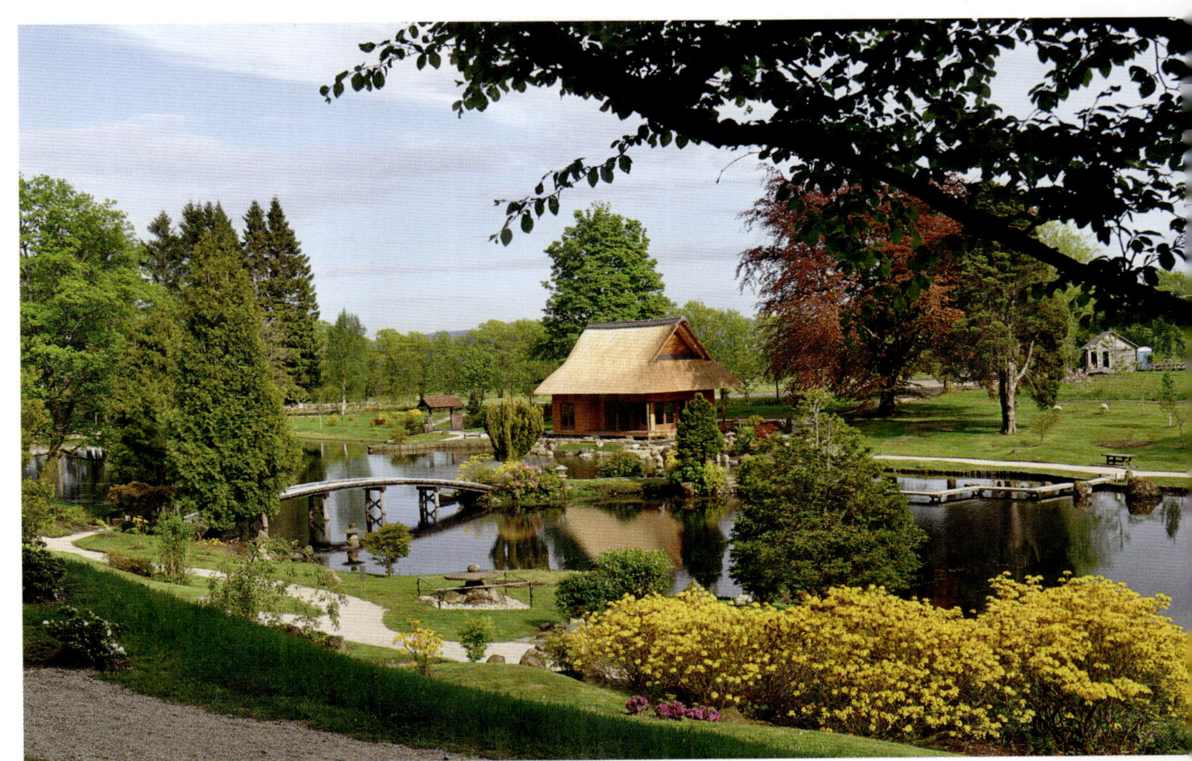

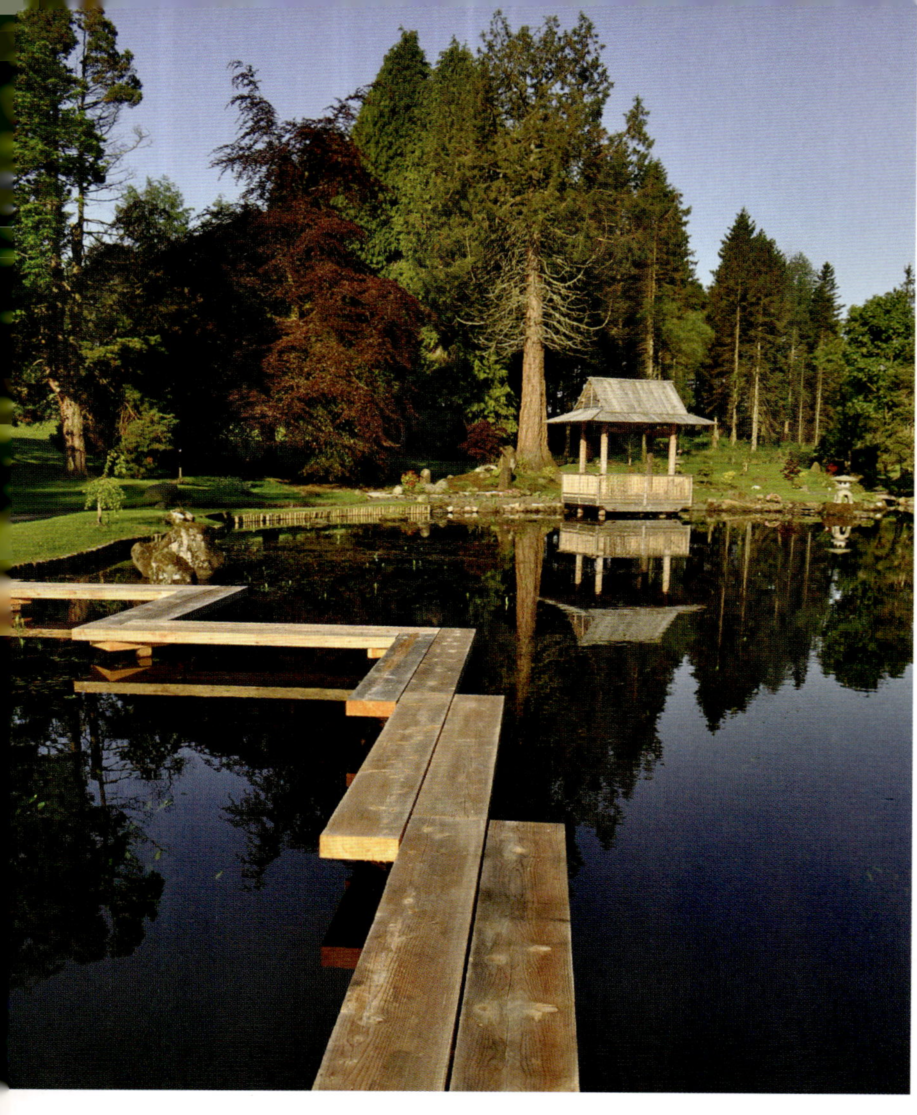

above left
Having strolled past the Garden Pavilion and under a *roji* gate, you reach the *yatsuhashi* bridge, a low, zig-zag pontoon of eight sections representing the twists and turns of life. The crossing, with submerged stones glimpsed through the water below, symbolises a walk through the galaxy.

below left
Having crossed the bridge of the Present, you arrive on the *nakanoshima*, the middle or sacred island, representing the state of limbo as the island of perpetual youth.

right above
Next you cross the *taikobashi*, or Drum Bridge, to the Afterlife (heaven) on the other side of the lake. The arch of the bridge makes it impossible to see at the start where it will land you. The slope rising up above the bridge to the north represents the foothills of Mount Fuji, one of Japan's most sacred mountains. Near the top is a circular seat under a rotating thatched umbrella, the latter a replacement of Ella's sunshade, where she would retire to read and think.

right below
Further up the hill on its new site beside the summer house stands a statue of a *tanuki*, or racoon, symbolising hospitality and safeguarding. Rosy-cheeked to show happiness, wide-eyed for good vision and with large testicles to indicate fertility, he wears a hat to keep off the rain and carries an accounts book and flasks of honey and *sake* (Japanese rice wine) for nourishment.

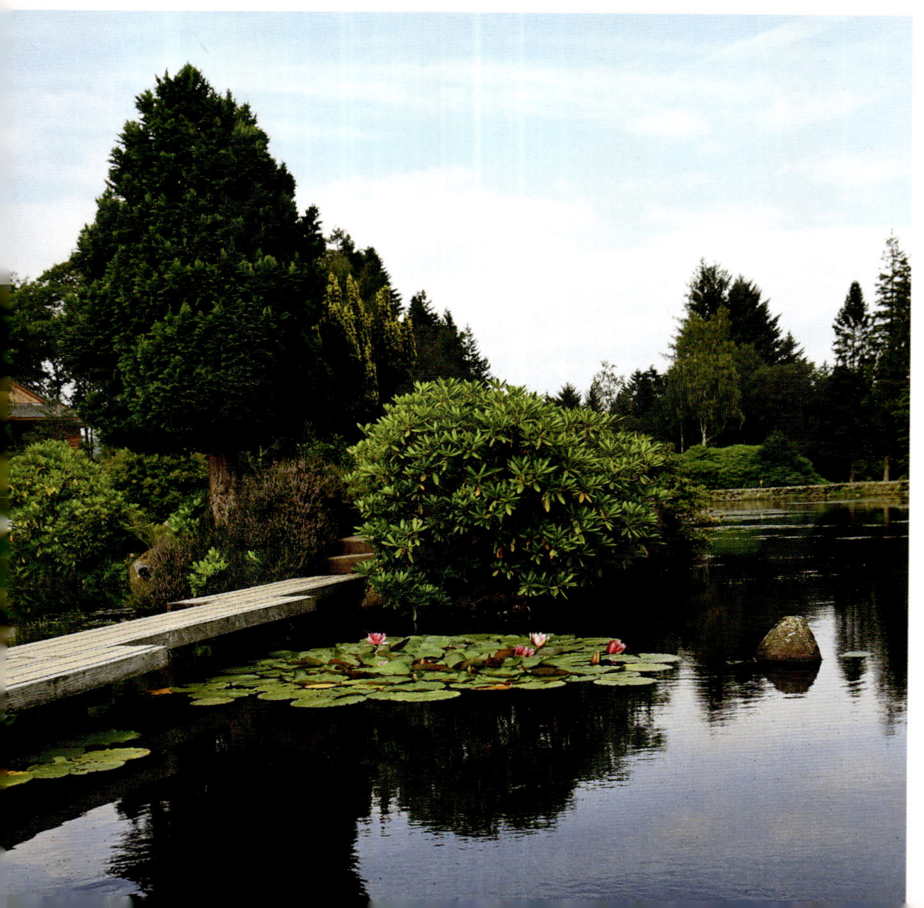

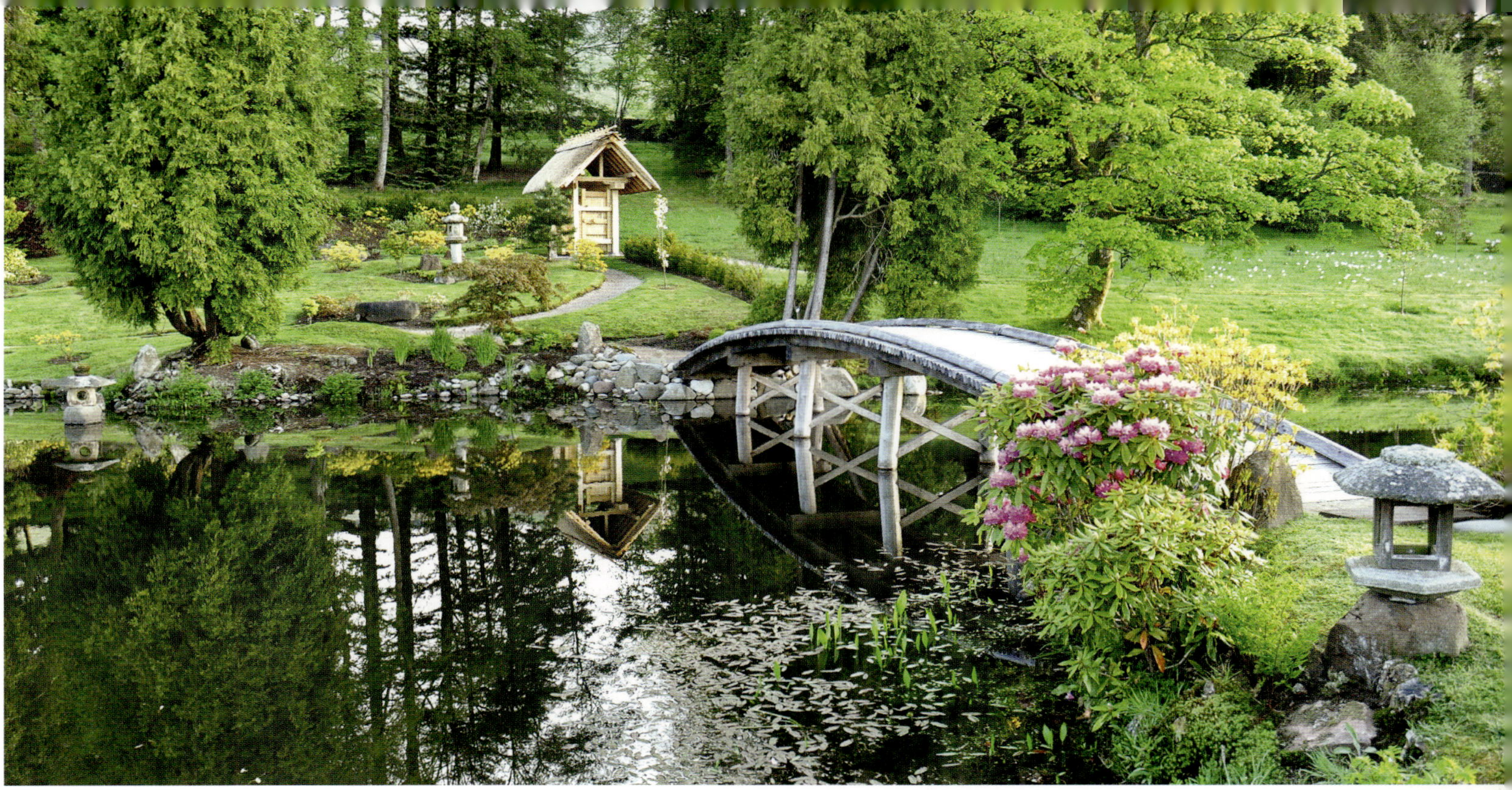

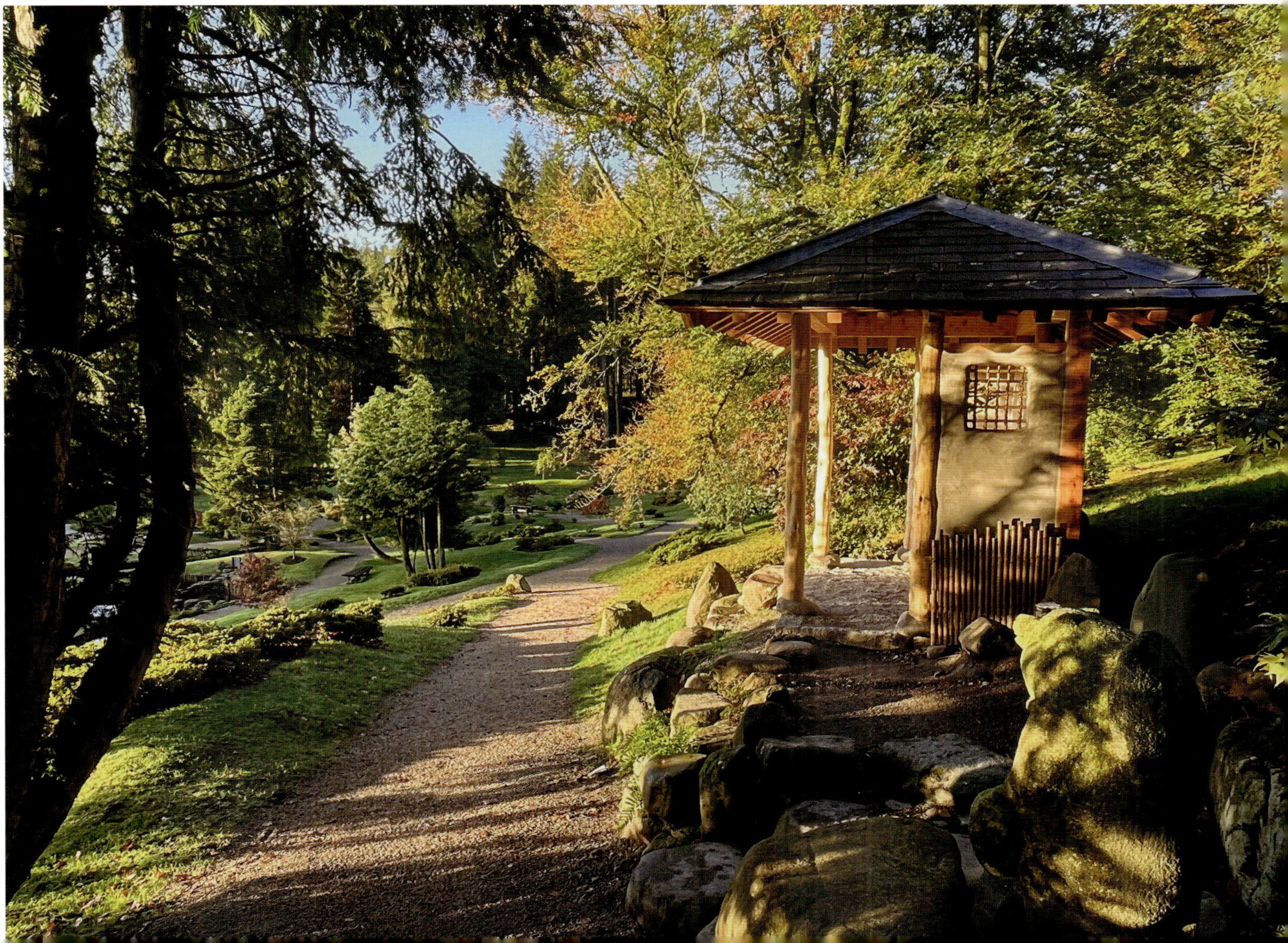

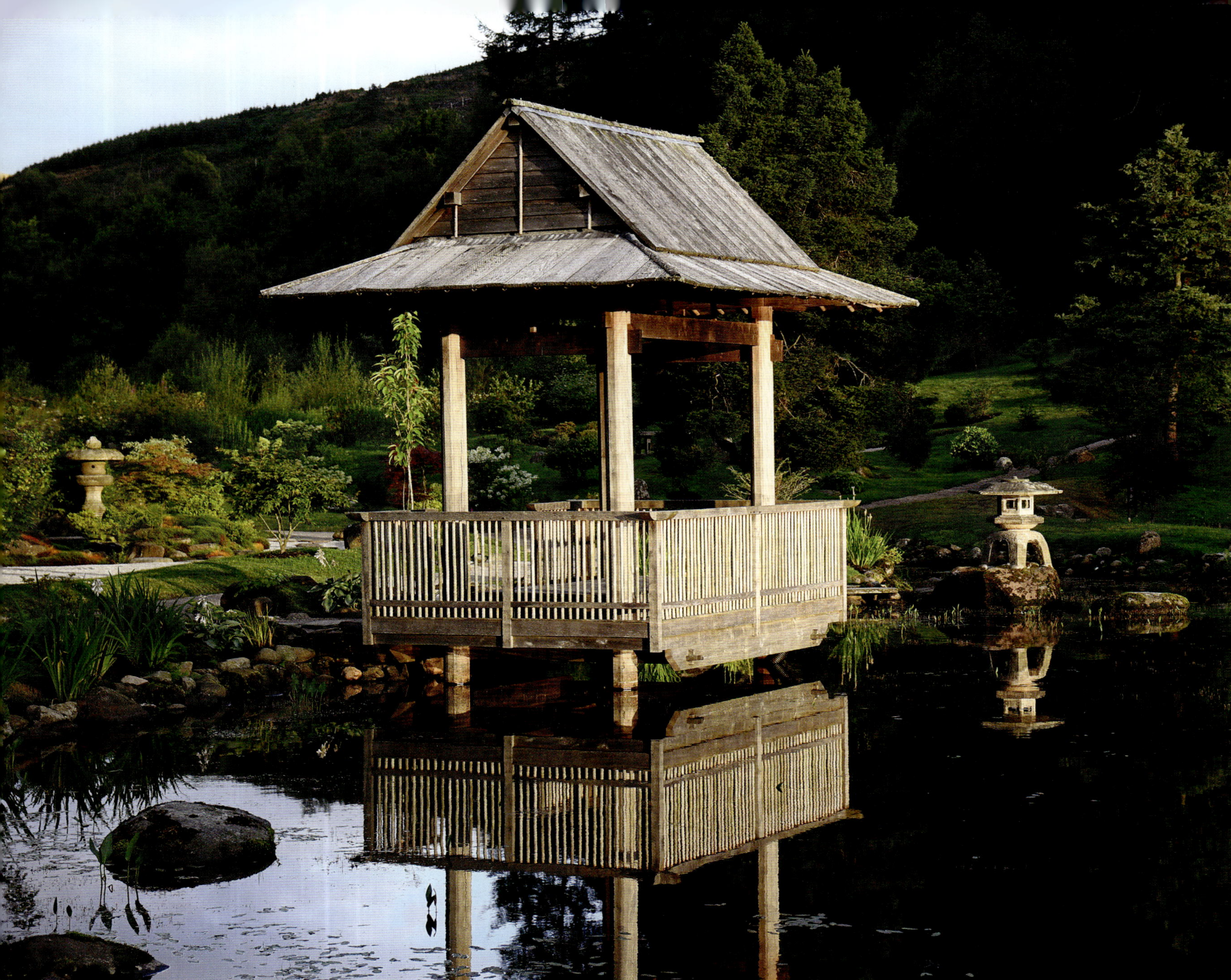

If you choose not to head for the Afterlife, but to
continue straight down the path, you come to an
azumaya, a summer pavilion of the type found in
many formal Japanese gardens. It is built in the
traditional style, open-sided and stilted out into
the lake so that occupants feel they are sitting on

the water.

below

Next to the *azumaya* is the *karesansui*, or dry garden, in which stones are carefully placed in a bed of gravel raked with ripple patterns, here symbolising water from the 'borrowed landscape' of the Ochil Hills behind. Dry gardens can be very sparse and quite large, with no clear explanation as to their meaning, leaving it up to the viewer to interpret as they wish. One of the most famous examples is at the Ryonan-ji temple in Kyoto, where 15 rocks are positioned in an expanse of gravel and a small amount of moss. Monks tidy away the leaves and garden detritus and rake the gravel daily as a form of meditation, often creating complicated patterns of circles and squares. In the *karesansui* at Cowden as developed since Ella's day, the meaning is clearer. The four moss mounds represent islands. On the central mound, the tall stone represents Sir Robert Stewart, the stone to the left his wife Grizel and the one to the right his daughter Sara Stewart, the current owner of the garden. The stone on the lower mound to the east represents a turtle, the one on the mound to the west a crane, symbols of longevity in the Japanese tradition. Ella Christie is represented by the larger stone between the central mound and the crane. Further up the slope, behind the *karesansui*, the path leads to the steps approaching the Inari Shrine.

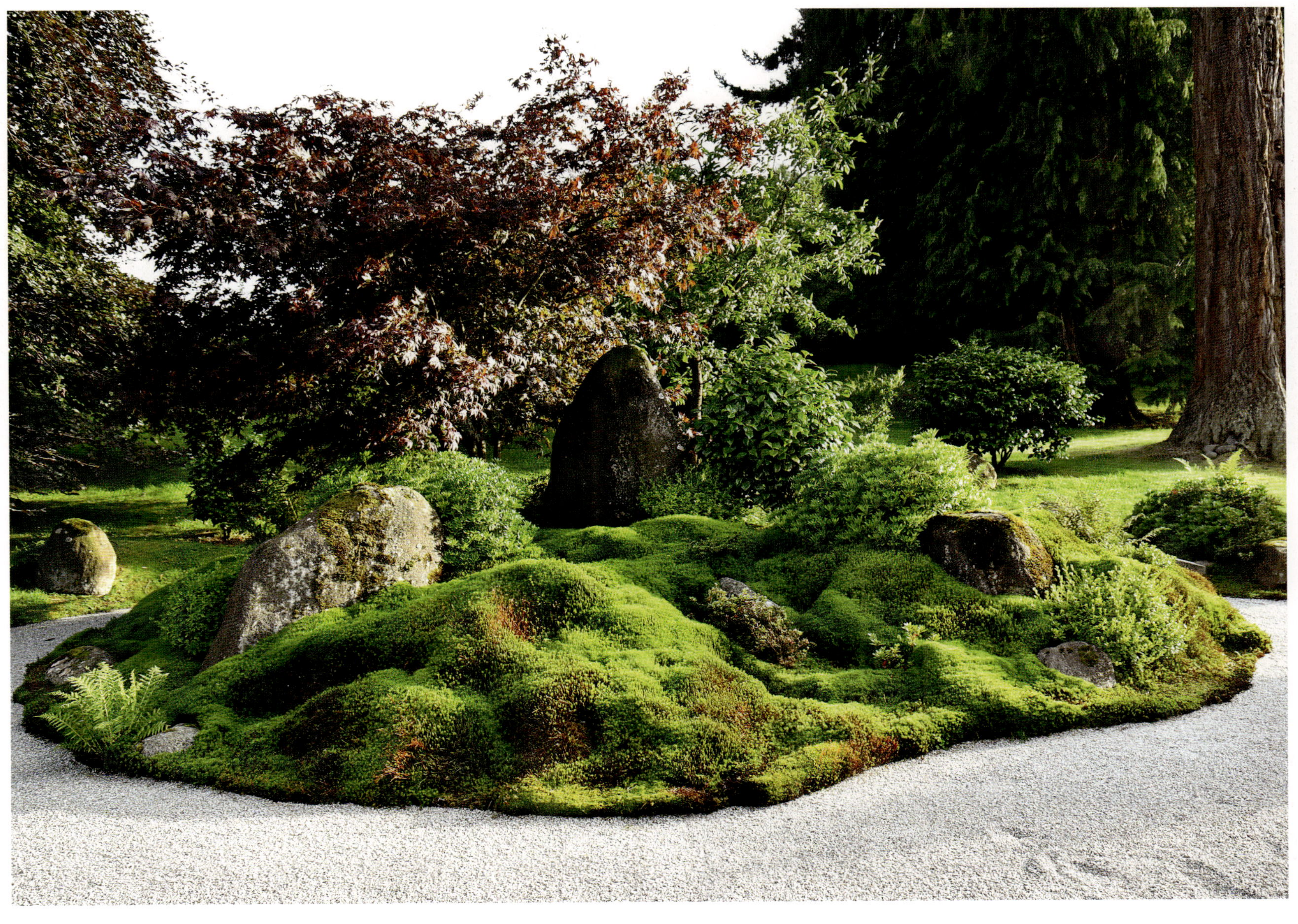

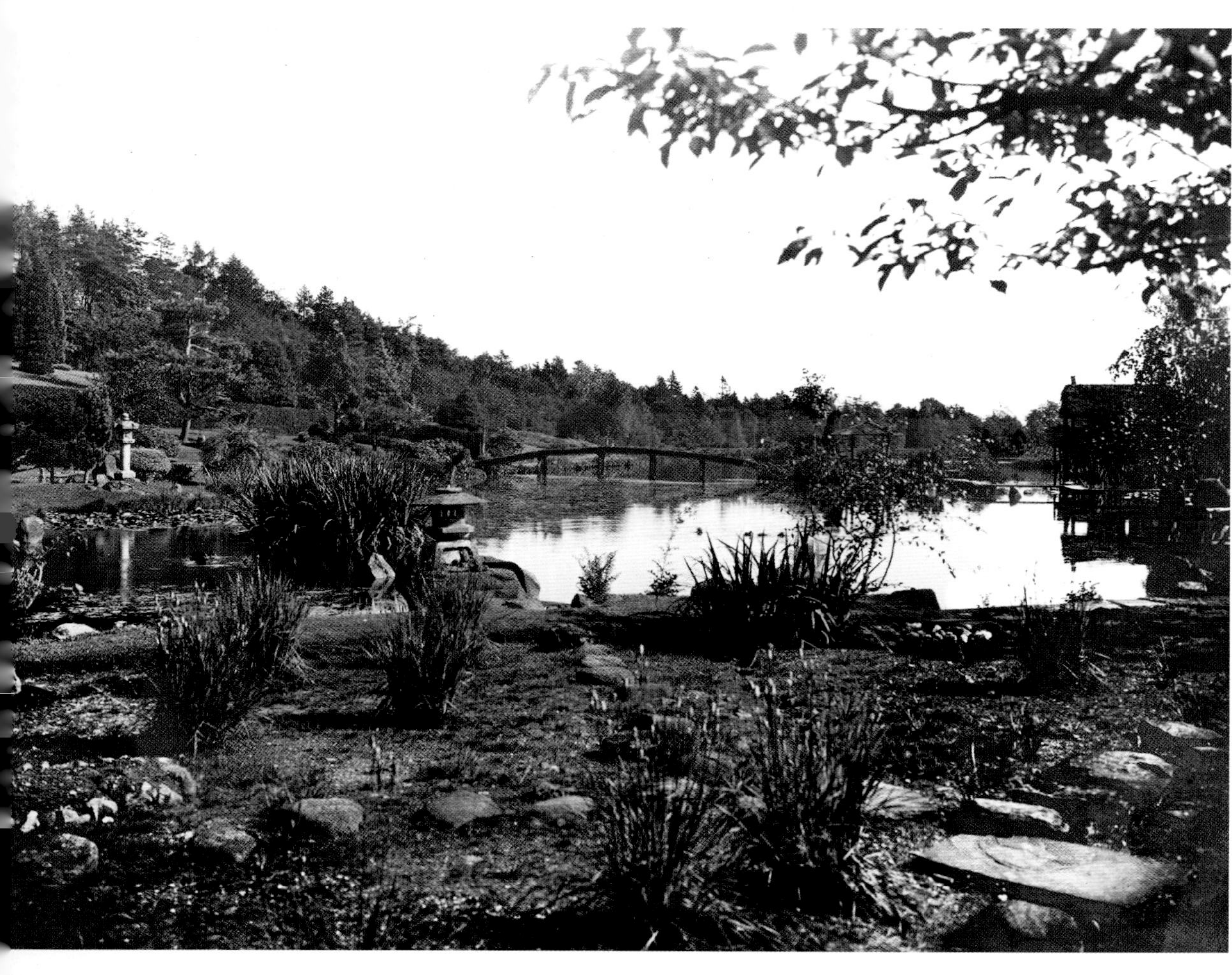

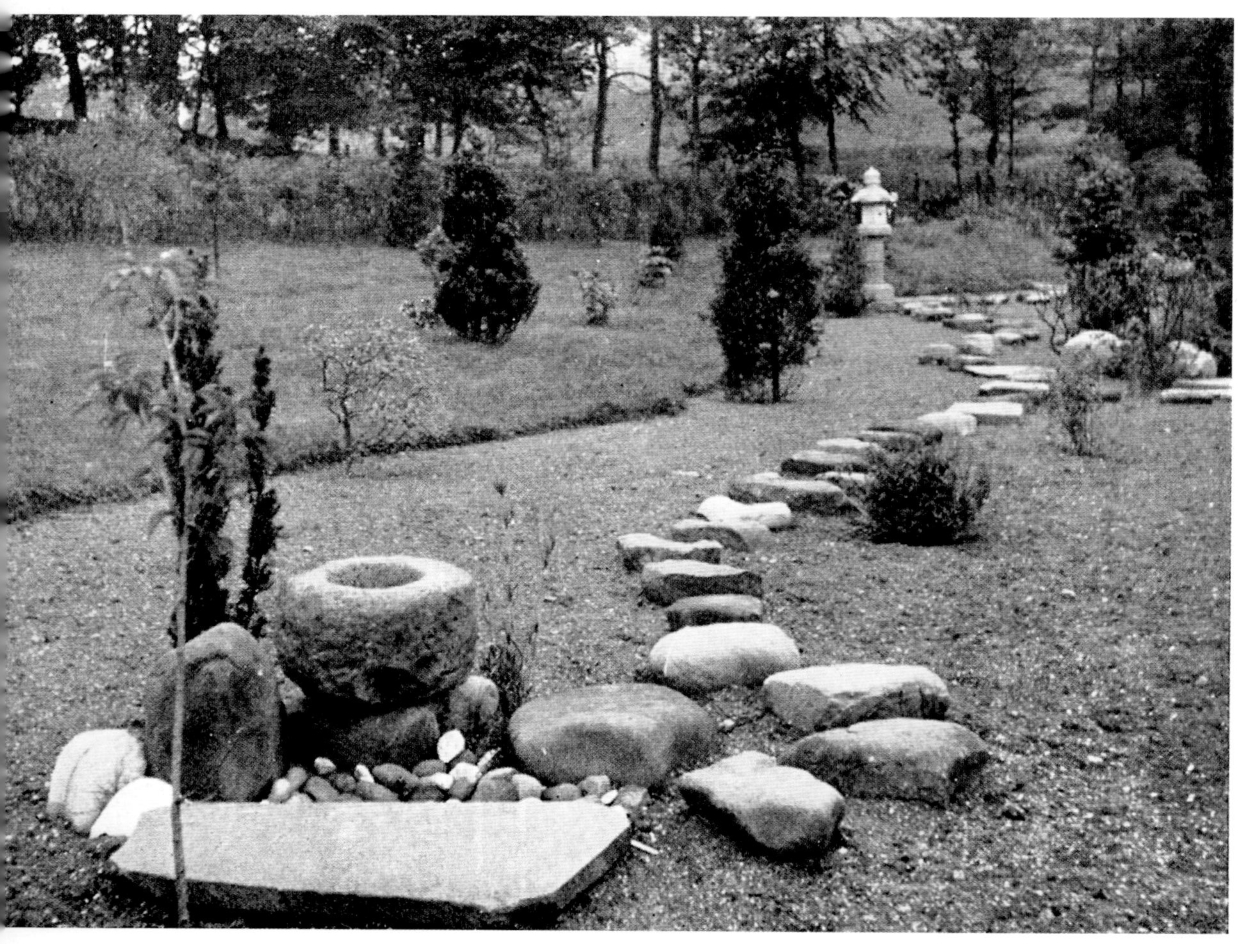

above and below left
These old photographs show the original appearance of the dry garden, with its hand-washing fountain (possibly originally part of the medieval castle), stepping stones and tall, hexagonal lantern or *kasuga dōrō*.

Many of the lanterns, such as the *yōsei dōro* seen in the middle of the upper photograph and the incomplete structure in the foreground, were pieced together from original fragments found at the bottom of the lake after the garden was vandalised in the 1960s. The lower photo shows the lantern on the *nakanoshima*, similarly made from bits retrieved from the lake.

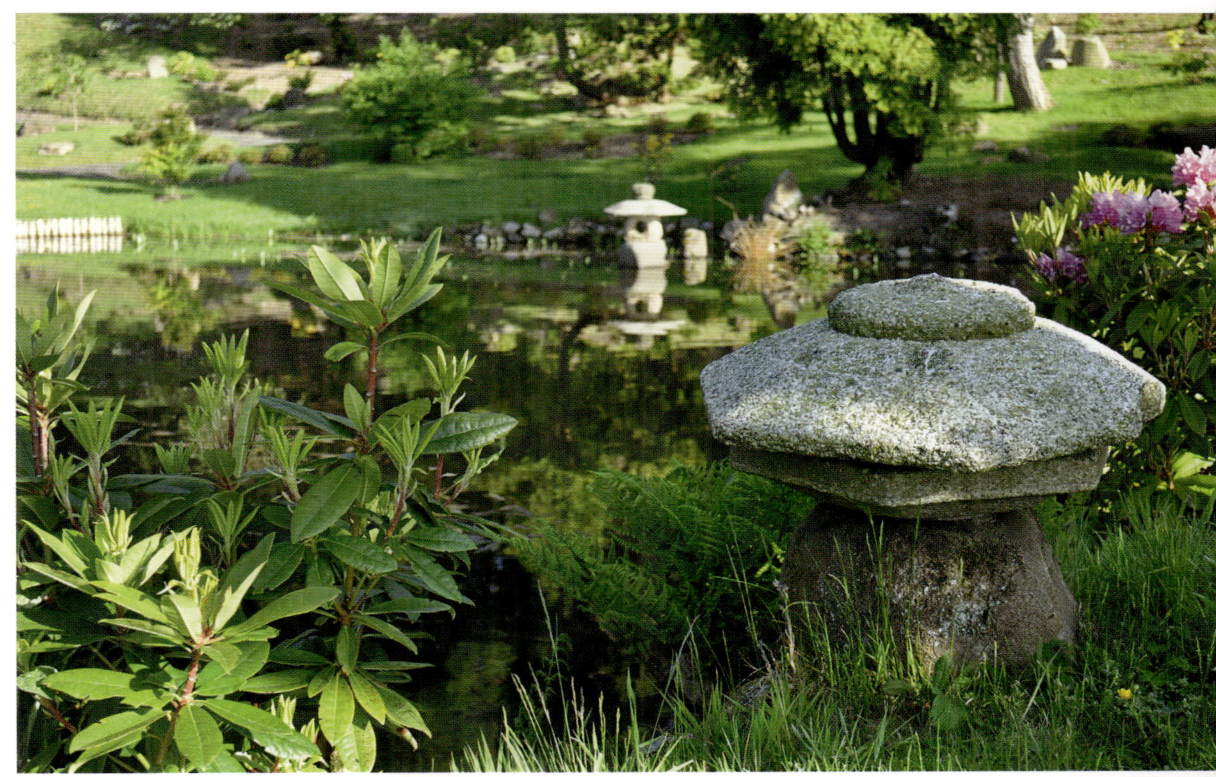

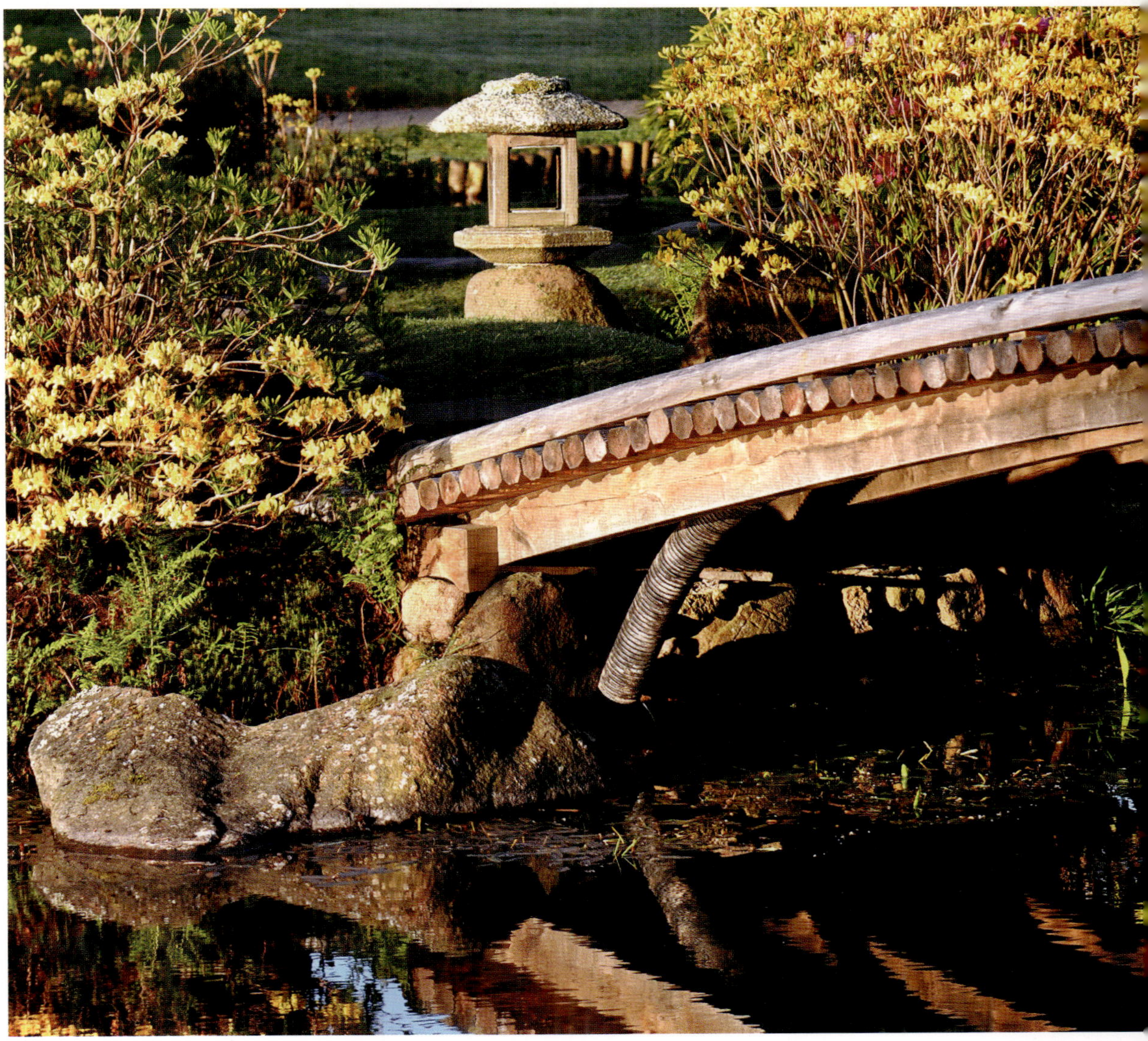

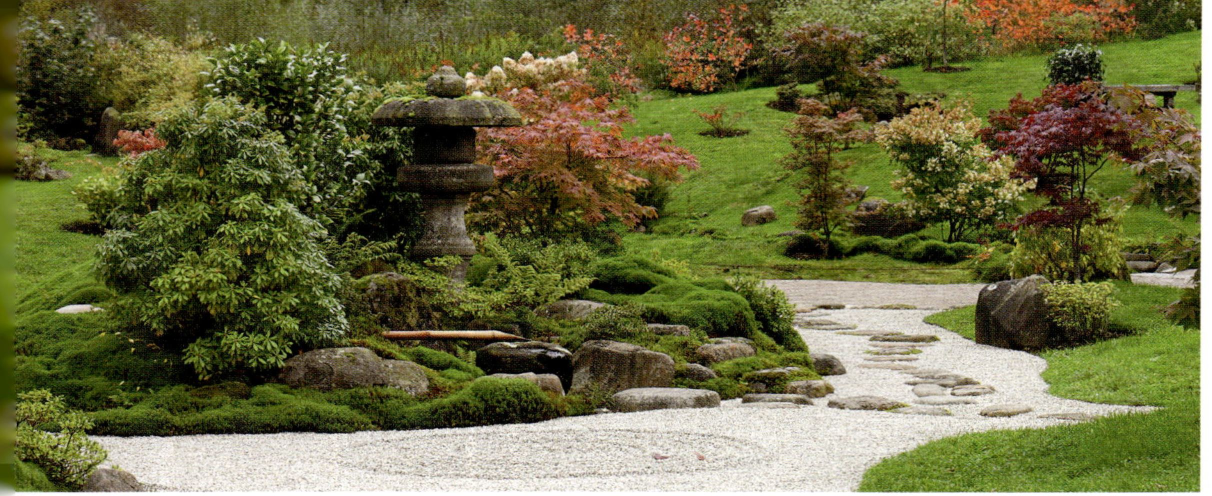

The *maru dōro* in the dry garden is surrounded by the luxuriant mosses that grow so well in this valley. Opposite, at the west end of the lake, is a *yukimi dōro*, a lantern standing on legs that is often put by water. The legs of this lantern are from a late 18th-century one from Nara that was among several antiques Ella had shipped over from Japan. It has been slightly moved from its original position.

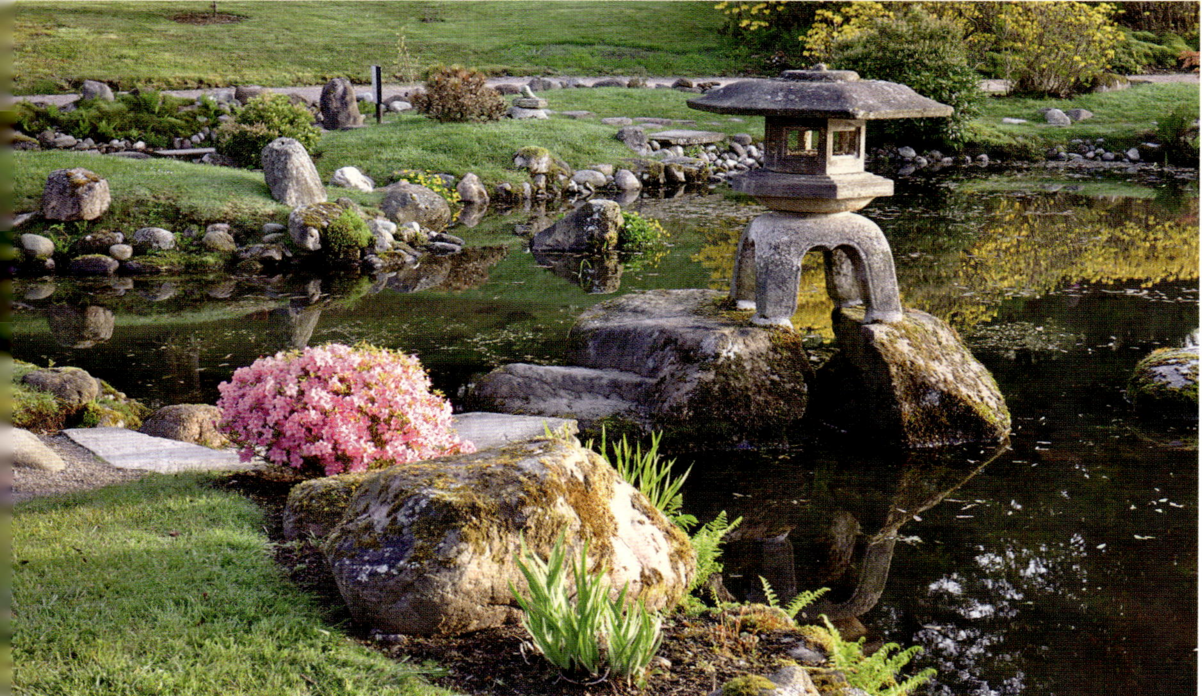

below left

Just below the North Gate is a *kasuga dōro*, generally a tall lantern with a square or hexagonal firebox often decorated with deer. The main column and base are original, the top section and firebox – or *hibukuro*, literally 'bag for light' – are replacements. The lantern has been moved about over the years, just as many of them were during Ella's lifetime.

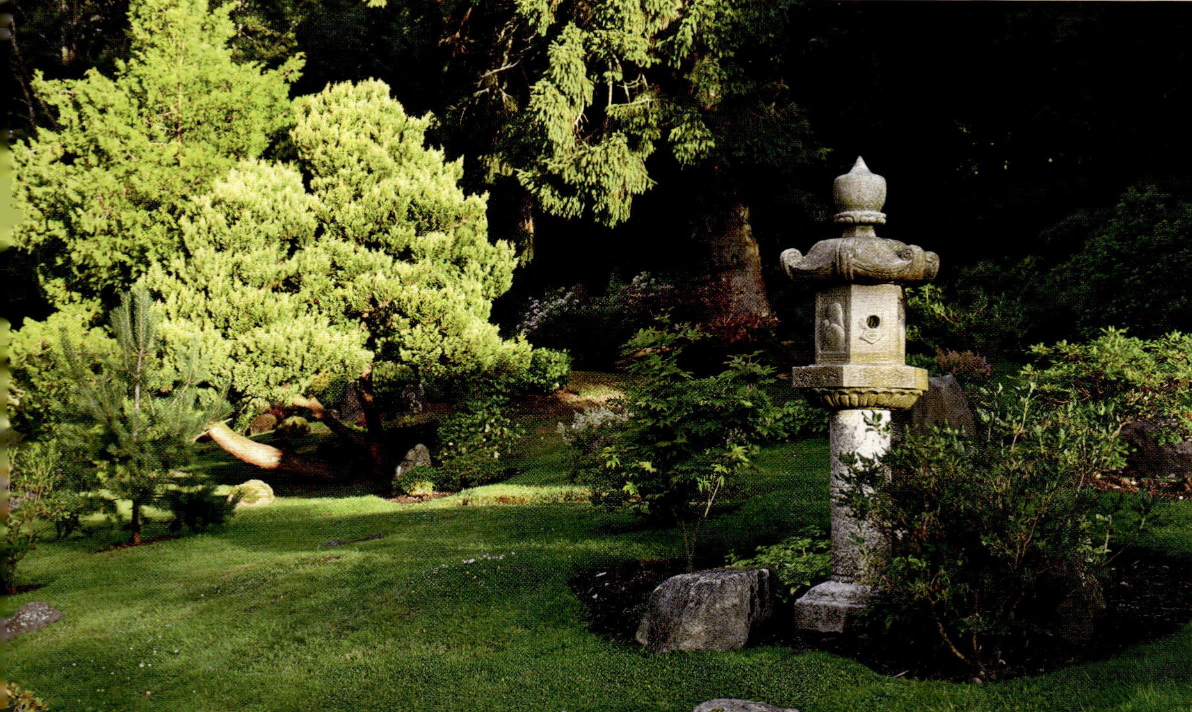

right above and below

These two lanterns – the first behind the *karesansui* at the bottom of the large redwood, the second on the Master's Isle – are made up of original parts that were discovered when the pond was dredged at the beginning of the restoration process.

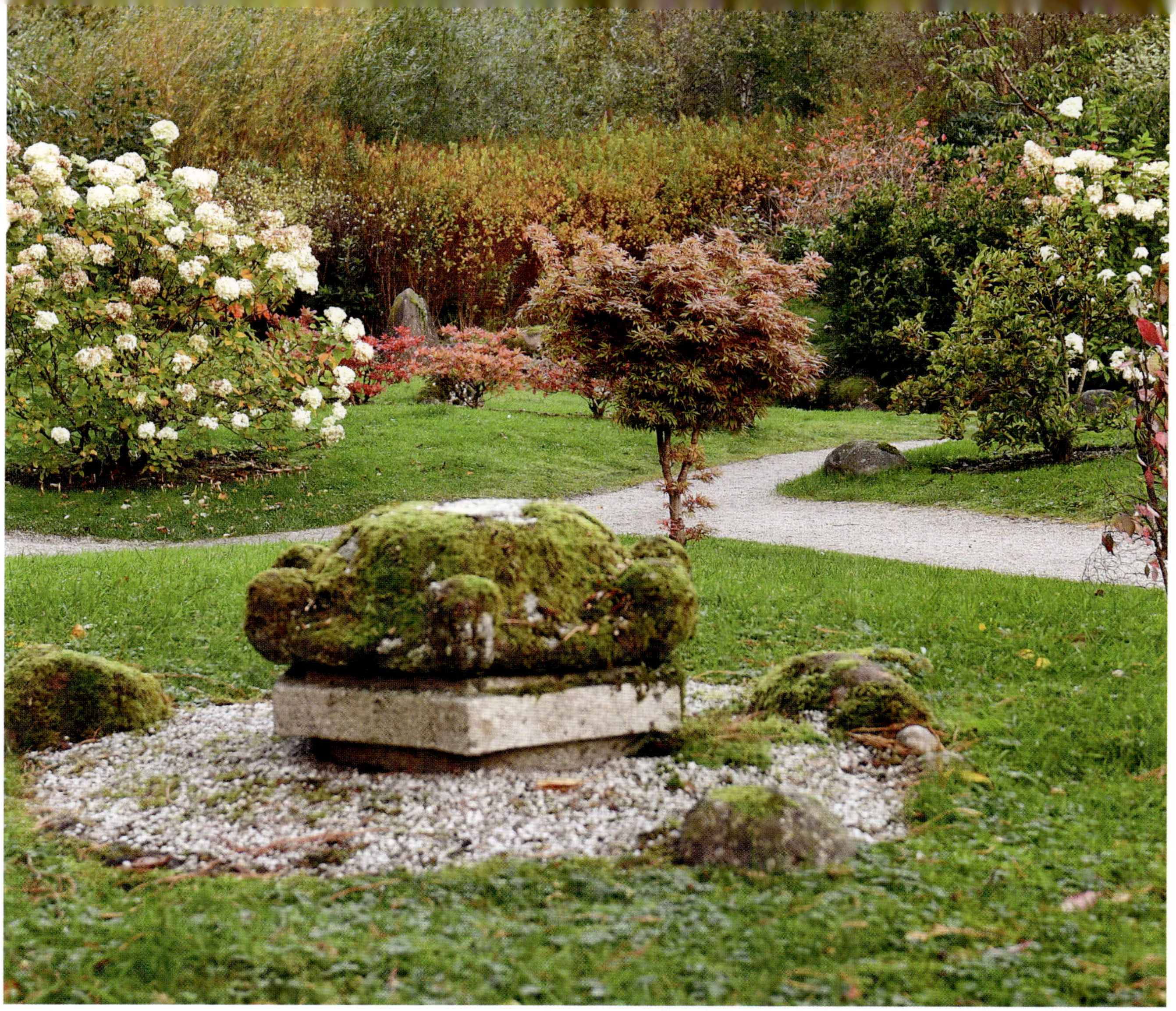

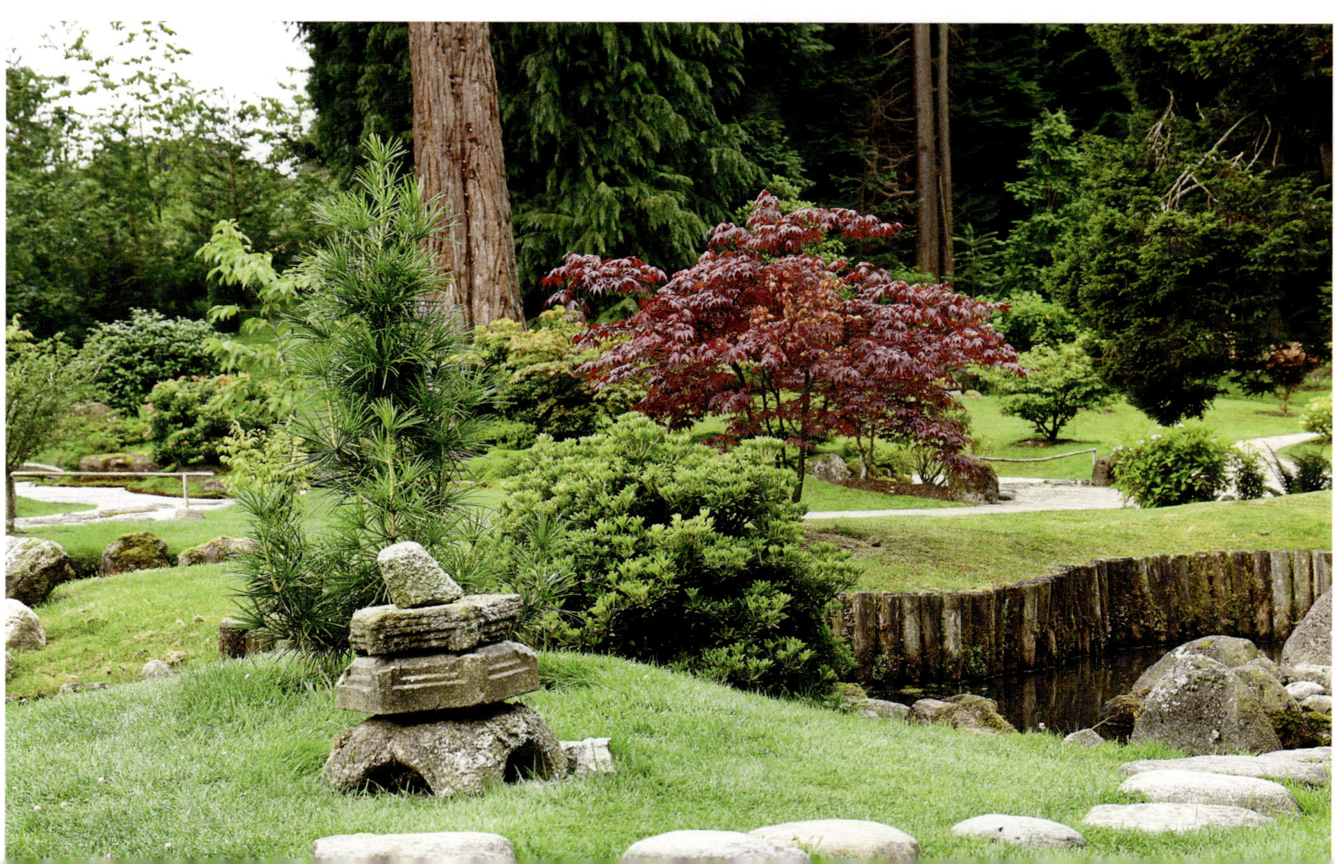

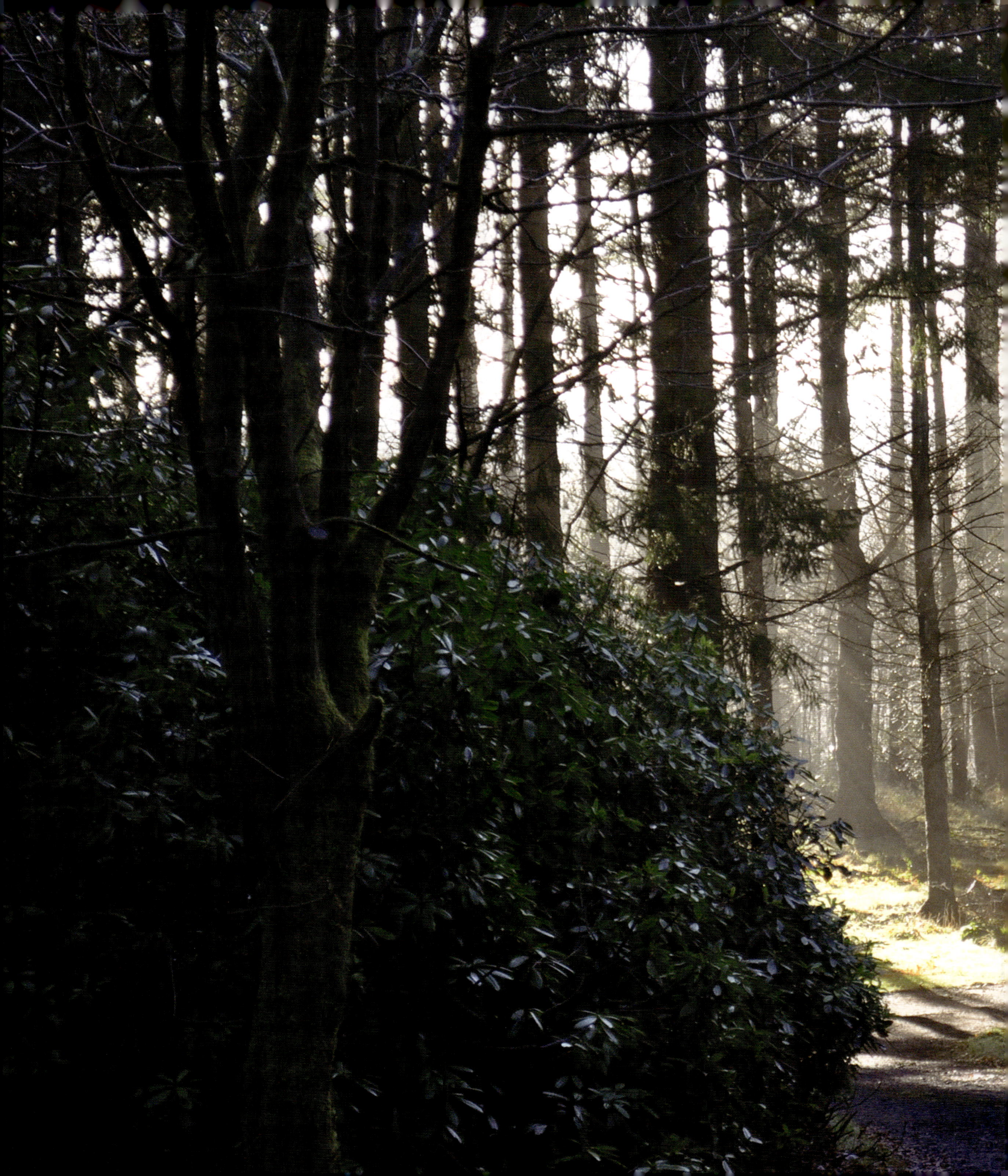

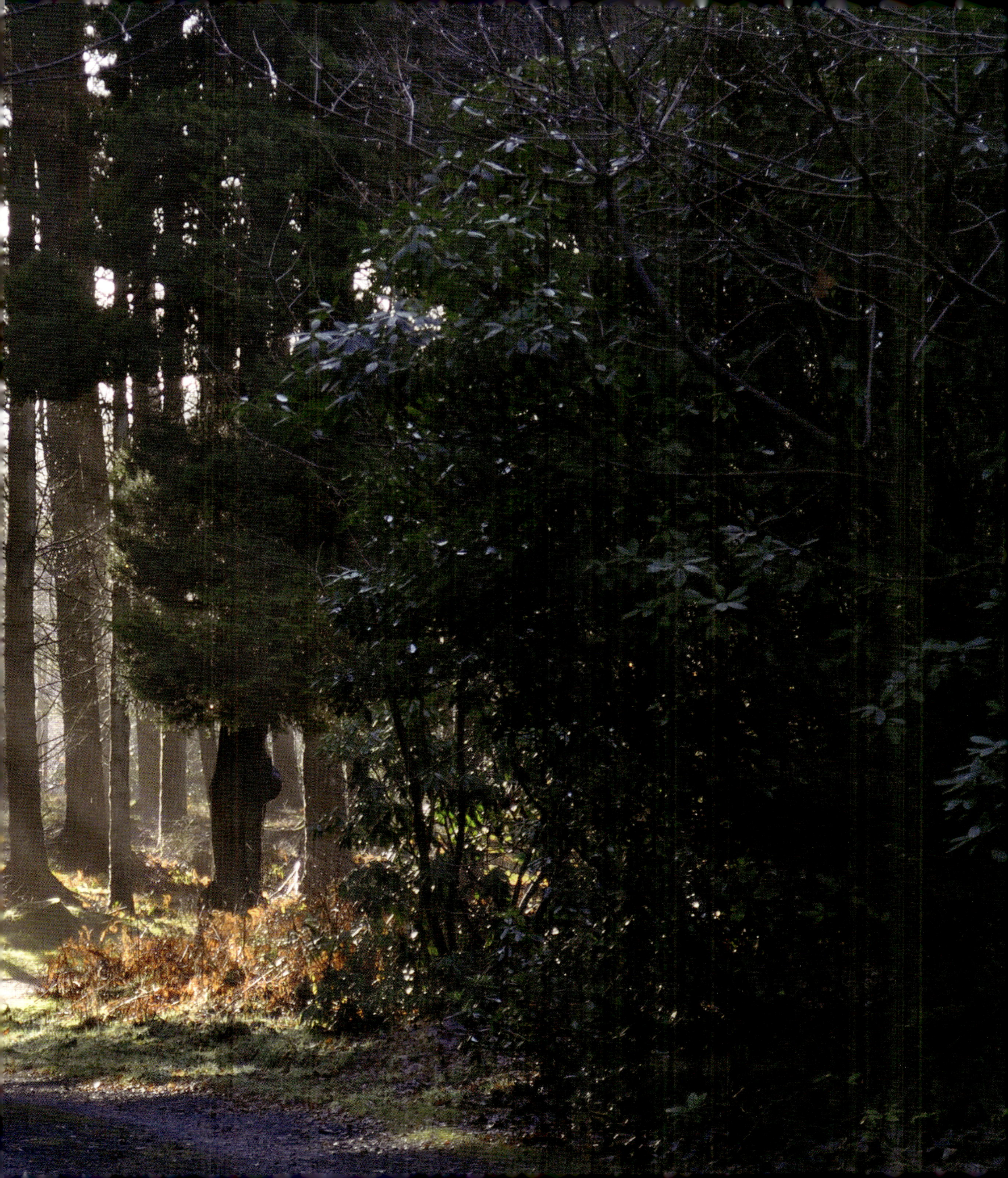

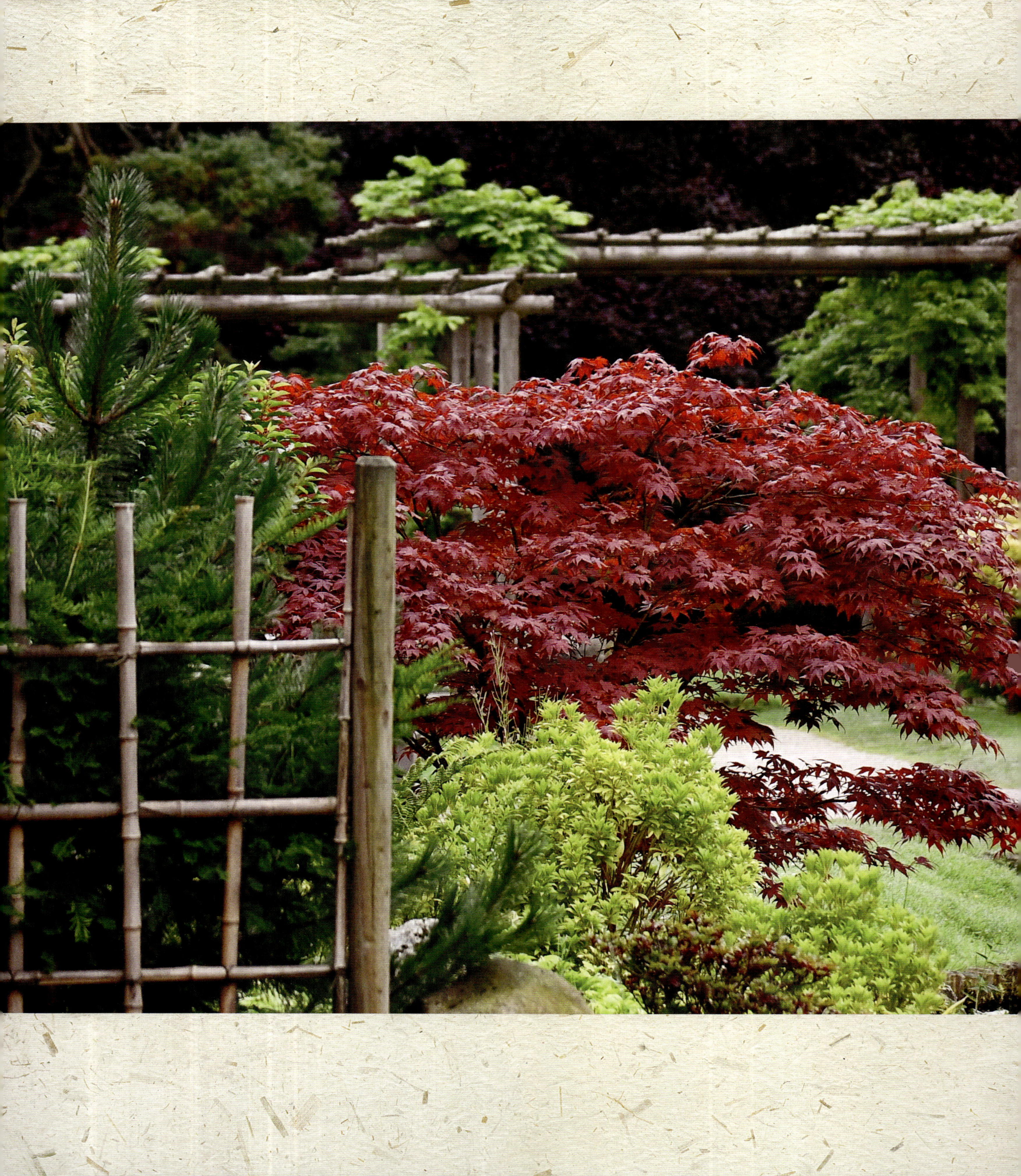

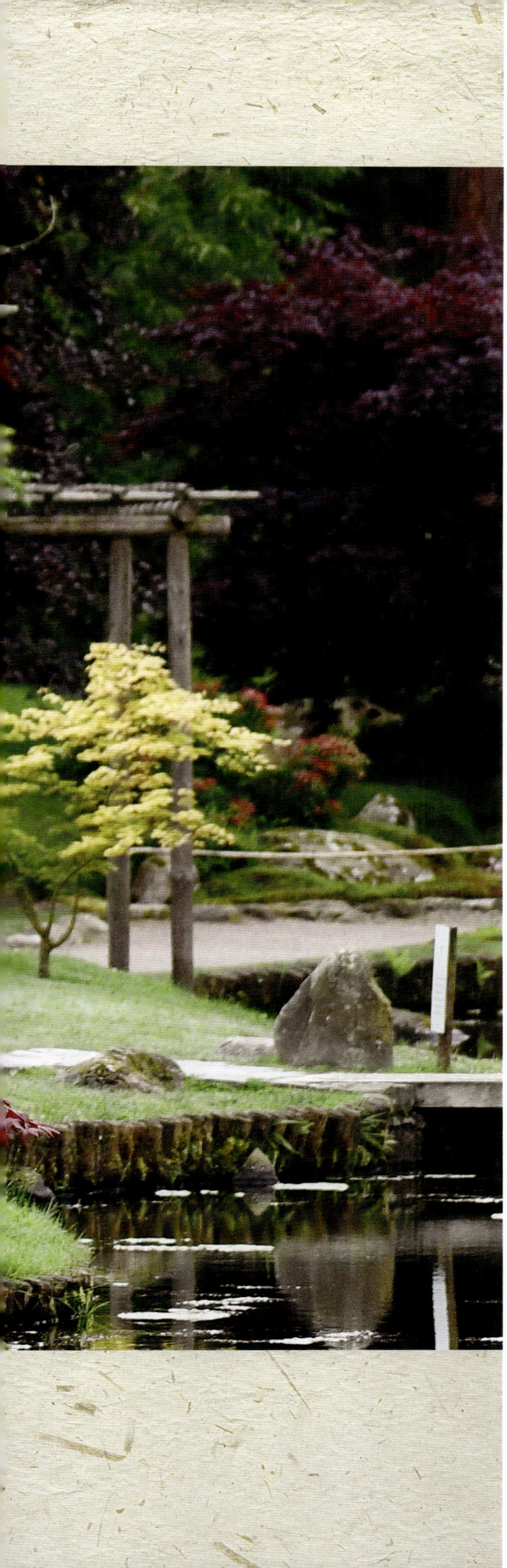

previous pages The back drive leading from the main road to Cowden Castle.

left The contrasting trees and simple bamboo structures enhance the overall harmony of the garden design with rich colour, texture and balance.

10 · Loss and Restoration

WHEN ELLA DIED in 1949, she left the Cowden estate to her great-nephew, Robert (later Sir Robert, known as Bobby) Stewart KCVO CBE (1926–2019), whose home was at Arndean two miles south of the garden. His father, Alexander (Bey) Stewart MC and bar, had been manager of Cowden Estate and Ella's heir, but he died in 1927 and the estate was put in the hands of trustees. By 1952, the castle, now riddled with dry rot, had become impossible to finance. It failed to sell for £1 and was demolished.

The garden continued to be maintained by the estate workers, but then, in 1963, vandals got in. They burnt down all the wooden structures, pushed the lanterns and rocks into the lake and caused much other destruction. With no funds available for restoration, gradually rhododendrons took over, trees self-seeded and it became impossible to see more than the perimeter path and the odd piece of rotting gate.

Although the garden that Bobby had known in its prime had largely disappeared from view, his emotional attachment to it remained strong. 'I remember spending most Sundays in the 1930s with [Ella] and my sister Grizel,' he recalled of his childhood visits. 'We would walk from the castle up the long, burnside

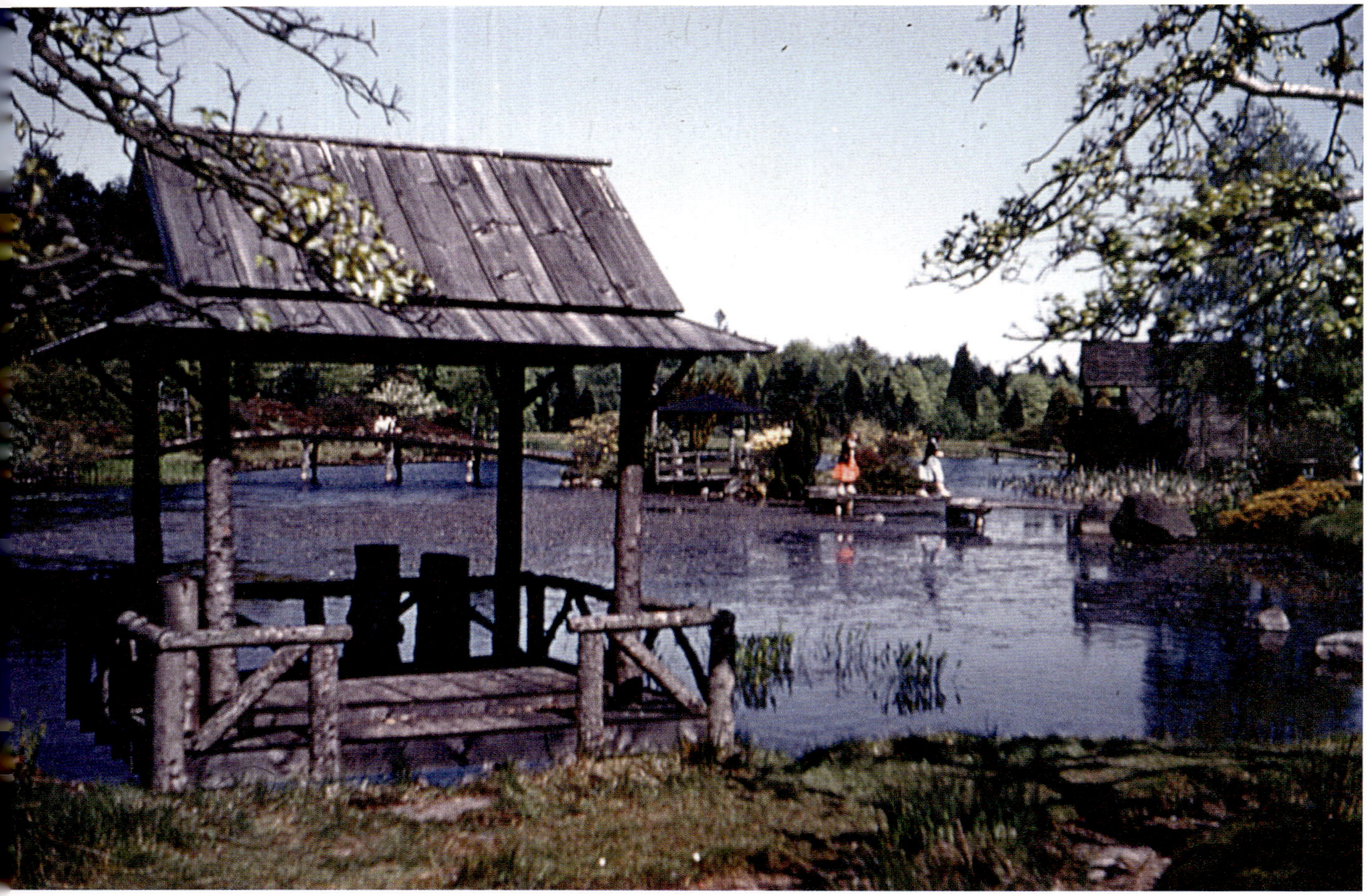

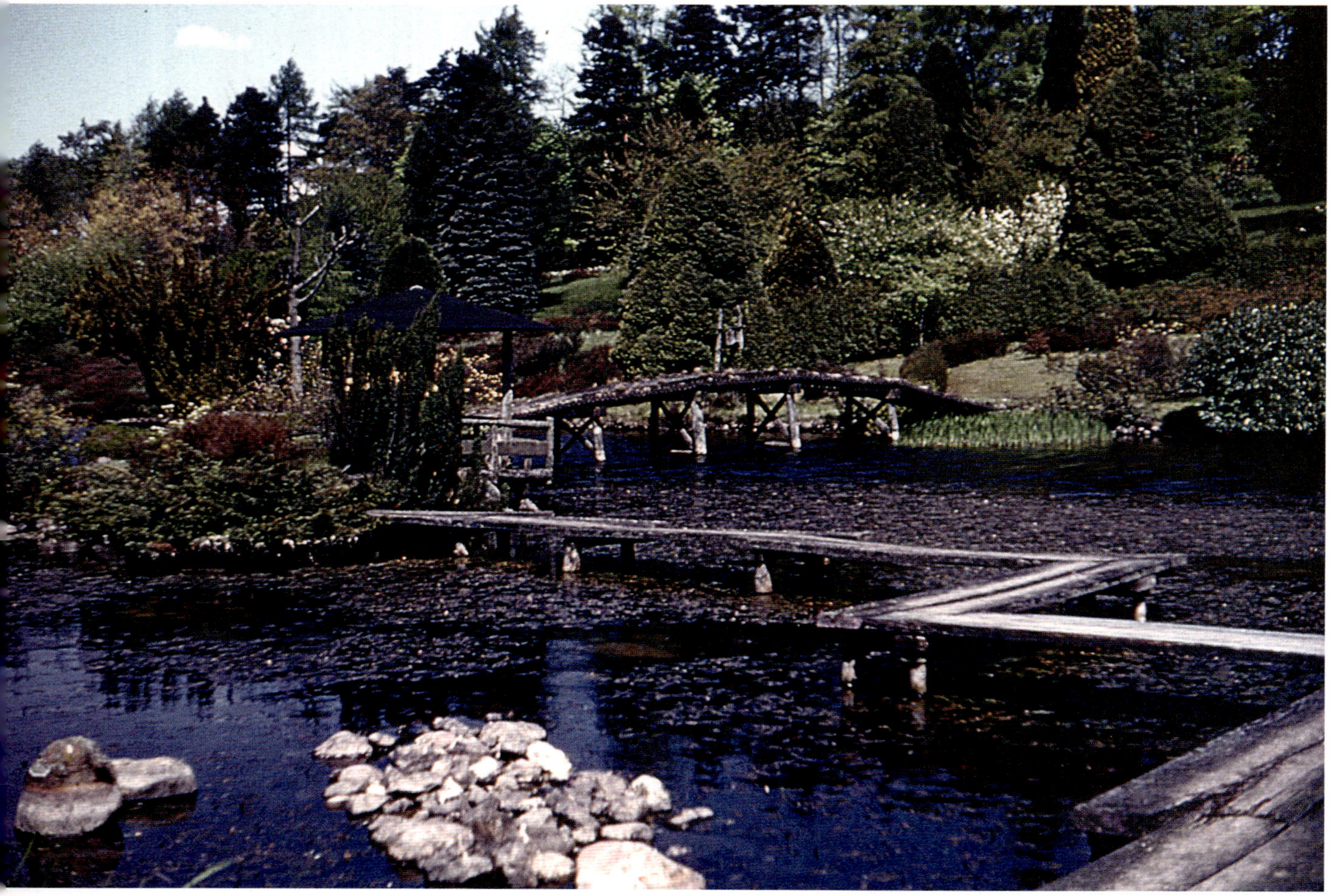

These views were taken in 1955, when the garden was opened in aid of St James's Church, Dollar. This was the last public opening before it was vandalised and became overgrown in the 1960s.

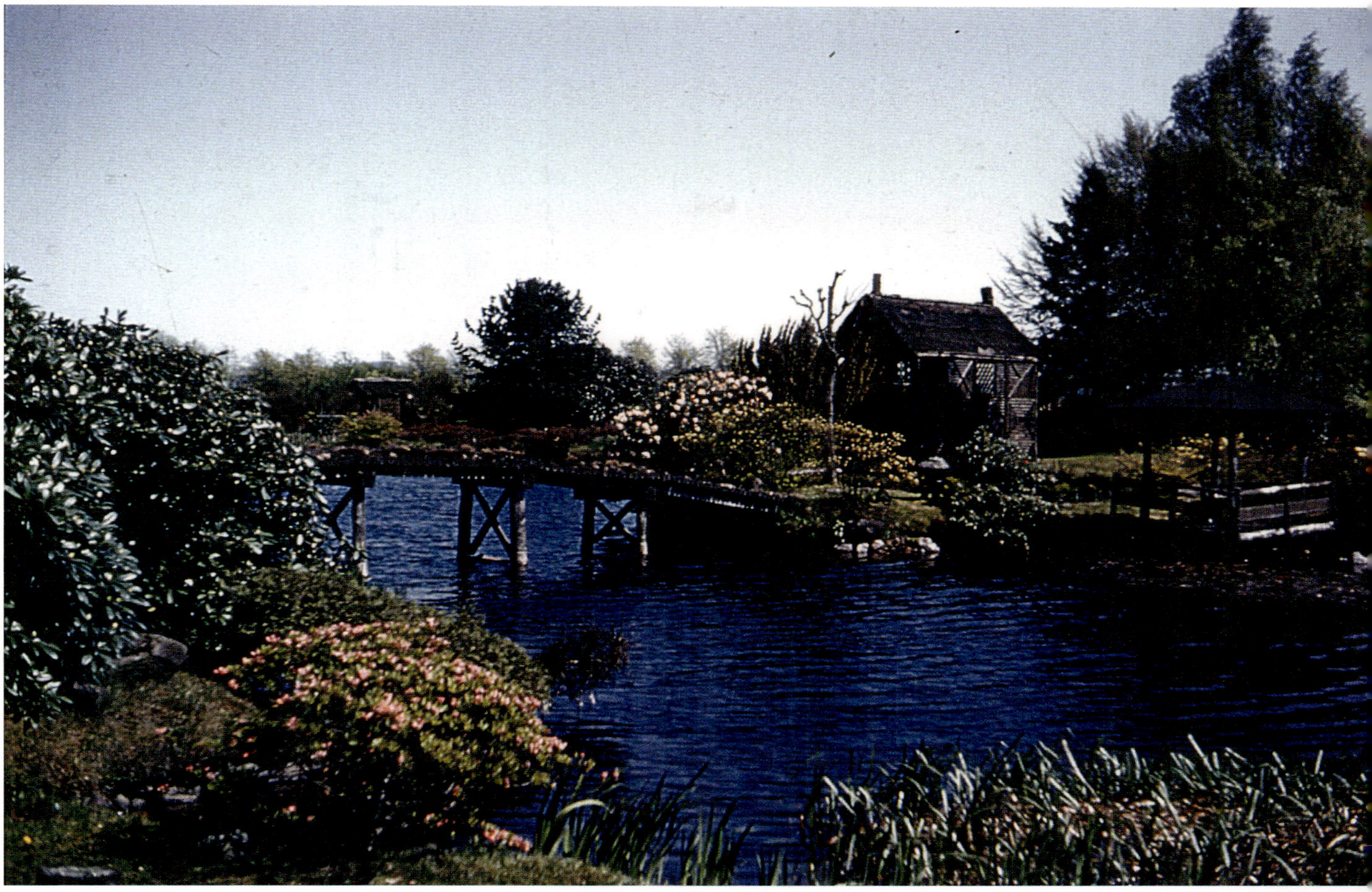

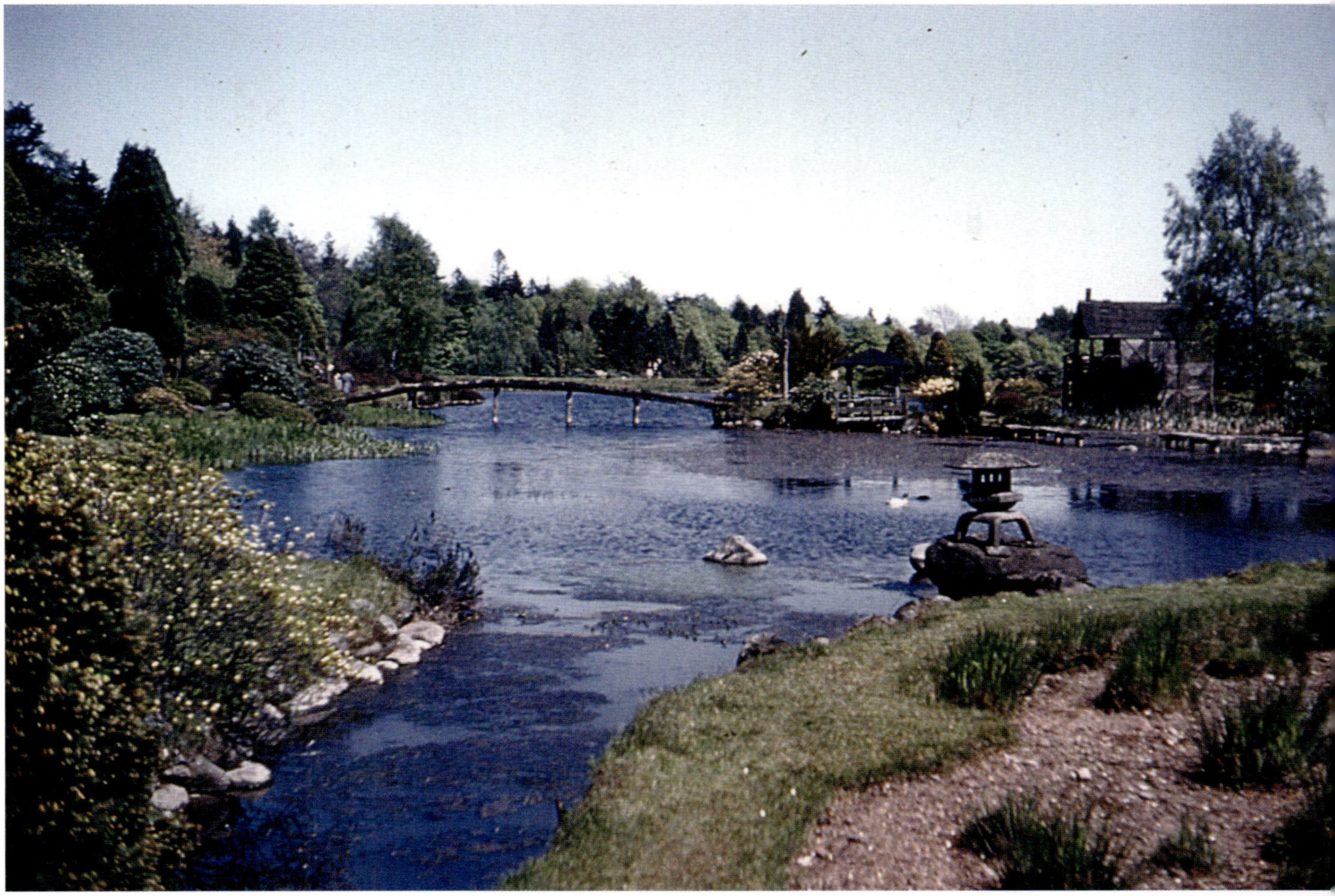

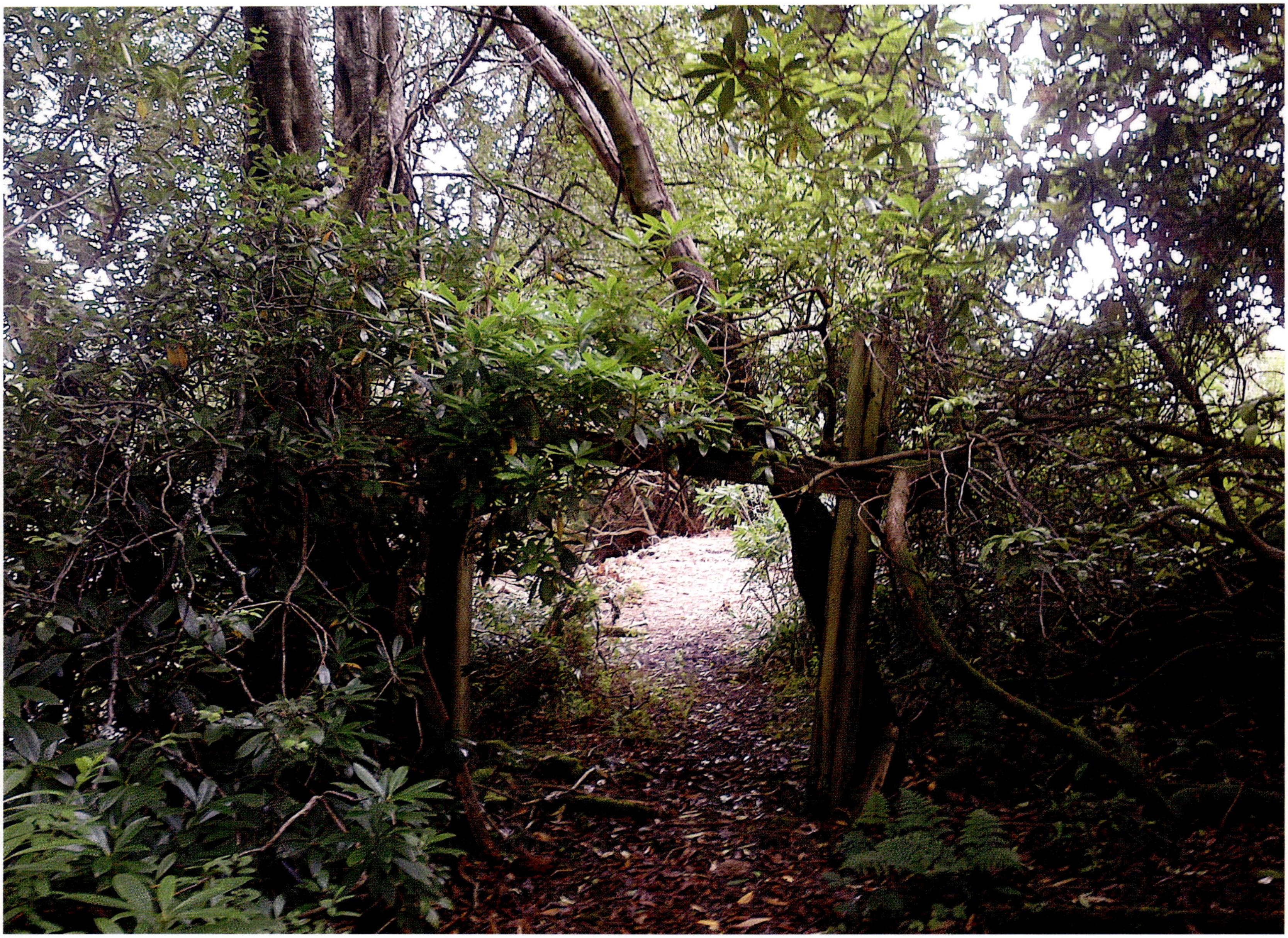

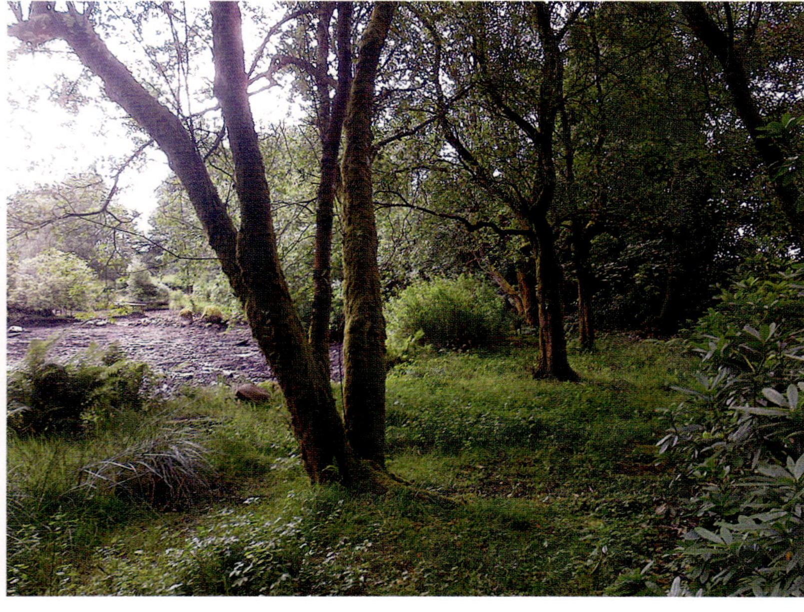
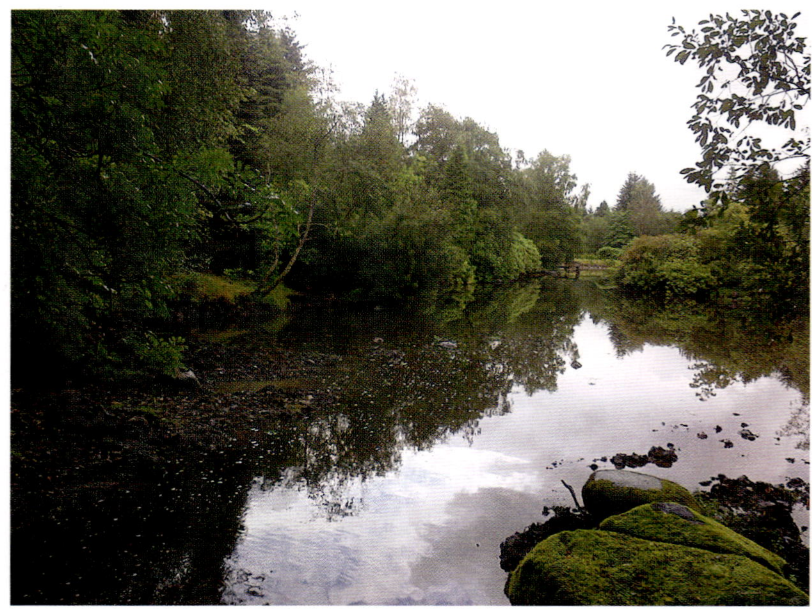

opposite Three photographs of the garden taken in 2012, before restoration. The upper one shows the remnants of the original Welcome Gate.

below Sir Robert Stewart and Flighty standing on the stone now occupied by the *yukimi* lantern. Sir Robert, Ella's great nephew, had inherited the Cowden estate on her death in 1949.

path, through the lime avenue to the garden, and have tea in the tearoom in the first floor of the boat house. The interior was furnished with symbolic wood carvings. From the balcony, one looked out onto the 'Slopes of Fuji' and the Ochil Hills in the distance.'[104]

Bobby continued to give talks about the garden and to take interested groups on visits, while heroically attempting to keep the choking overgrowth at bay. During the 1980s, a Japanese consortium approached him with plans for a golf-course development, and various other developers came up with proposals for tourism ventures, but none of these appeared to be sympathetic solutions for the site and they didn't appeal to him.

In 2012, he passed the ownership of the garden to his daughter Sara, who had spent many hours of her childhood helping him cut back the vegetation. Although Sara had only ever known it in wreckage, her great-great-aunt's creation had become a large part of her life and she was keen to see it returned to its former glory. It was a chance telephone call in 2013 that provided the impetus. She happened to be visiting her parents that day, when the telephone rang and she answered it. The caller was Professor Masao Fukuhara of Osaka University of Arts, who was visiting Scotland to give a lecture and wondered if he could pay a visit to see the world's only accredited Japanese garden designed by a woman. The Stewarts invited him over and he arrived with his two assistants, Ai Hishii and Junya Matsukawa. As Bobby and Sara showed them round, both were greatly taken by the Professor's hands-on interest and genuine enthusiasm. He had

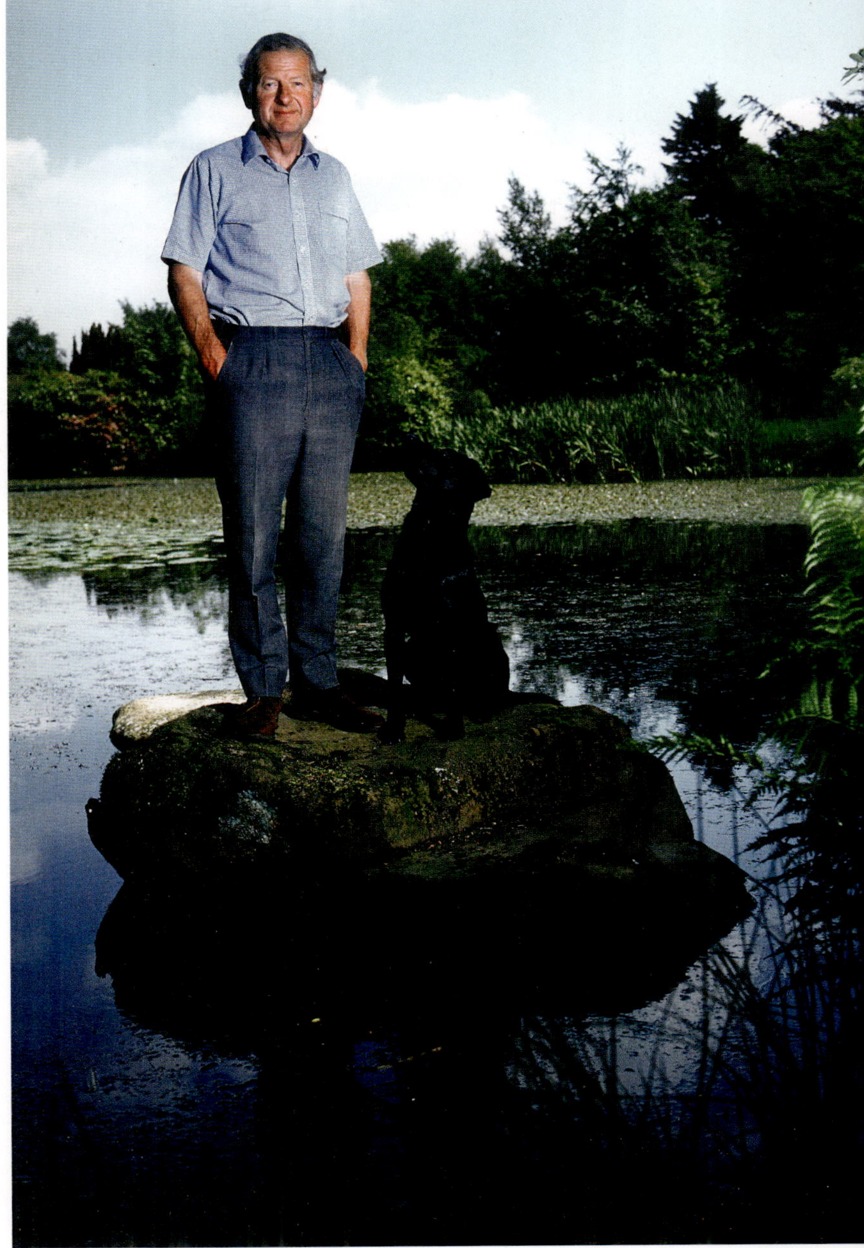

already made a name for himself in British gardening circles. In 1994–96, he had overseen the restoration of the Japanese garden at Kew and he was a gold-medal winner at the 2001 Chelsea Flower Show for what is now part of the National Botanic Garden of Wales. That year, he had also led the restoration of the Japanese garden at Tatton Park, Cheshire. The outcome of his visit to Cowden was that Sara appointed him to oversee a programme of restoration.

Work started in 2014 under a team brought over from Japan. Between then and March 2020, Fukuhara, Hishii and Matsukawa would come twice a year for 10 days at a time, tackling a different section of the garden on each visit. Often, they would bring a group of the Professor's students to give them valuable experience of working in a historic garden.

The project had not been going for more than six months when it became apparent that Fukuhara's original estimate of £50,000 was far too low. And so, Sara set up the Japanese Garden at Cowden Castle SCIO, a charity eligible for grants from foundations and other bodies that support the advancement of the arts, heritage, science and environmental protection and improvement.

The resulting funds enabled Fukuhara and his team to continue with the hard landscaping, the construction of an access, the design and erection of the *azumaya*, gates and other structures and the employment of a full-time gardener. None of the original lanterns survived complete, so Fukuhara sourced replacement parts of a similar vintage from Japan. The Garden Pavilion, designed by Ella's great-great nephew Hugh Shaw Stewart of Forster + Partners, was completed in 2022 and a replica of the revolving thatched umbrella known as Ella's Sunshade was installed in 2023.

The Garden Pavilion is a recent addition – an alternative to replacing the vandalised boathouse, which would have been too small to accommodate the number of visitors that the garden now attracts. Professor Fukuhara created the four mounds at the west end of the dry garden, an area for which there was no surviving photographic record. And now Ella and Bobby's vision for a dwelling overlooking the lake has been realised with the creation of Christie Lodge, a Japanese-style holiday cottage on the northeast slope.

Sharaku-En re-opened fully in 2019, and now has three full-time gardeners assisted by a team of volunteers, who continue the work of Ella, Taki Handa and Professor Fukuhara. It remains Britain's most authentic Japanese strolling garden, designed for aesthetic pleasure and as a place for reflection and meditation, a cultural bridge between two island nations. Ella's spirit has returned and we can picture her sitting up at the top of the slope as she loved to do, in her 'gazebo of eastern design built by her own foresters, watching the ripples on the lake'.[105]

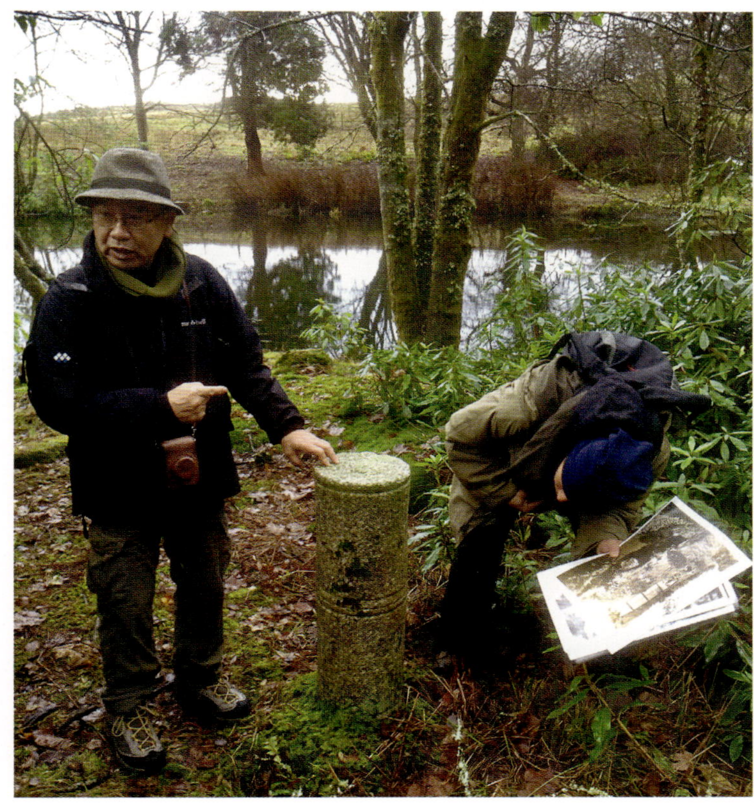

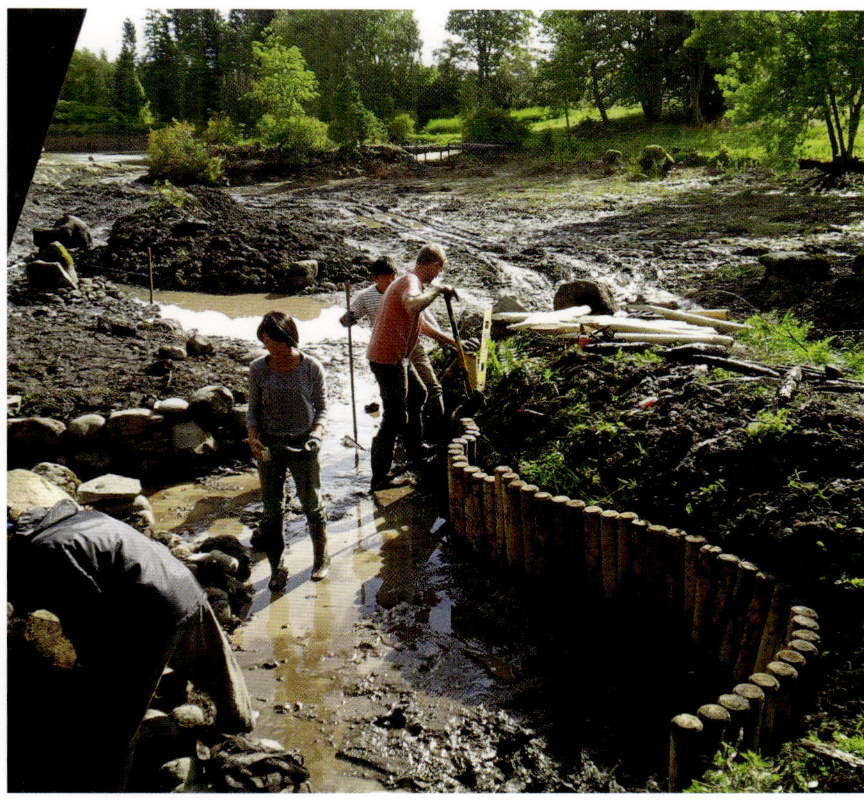

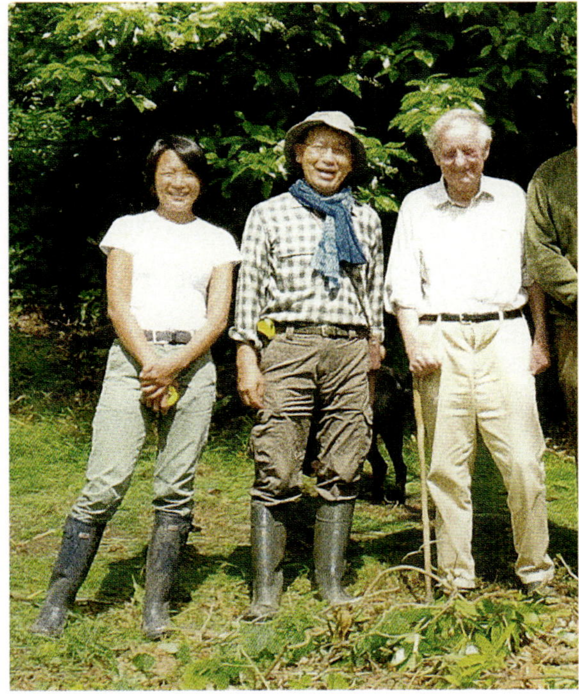

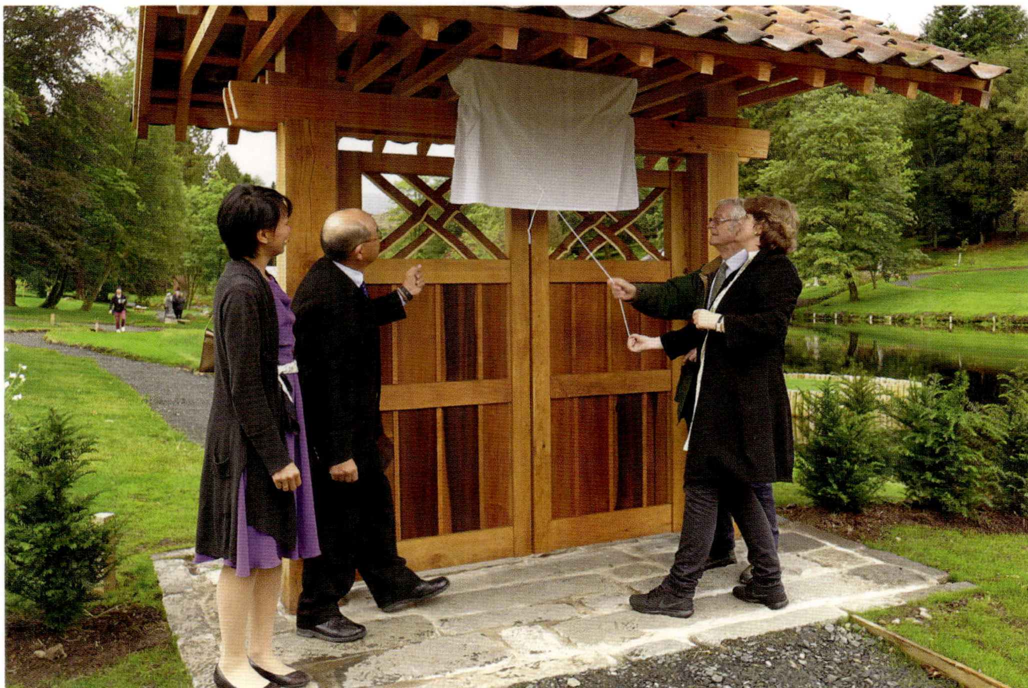

The Japanese Garden in 2025:
an aerial view of the lake looking
over the *azumaya* to the *yatsuhashi*
and *taikobashi* bridges with the
Garden Pavilion and Welcome
Gate beyond.

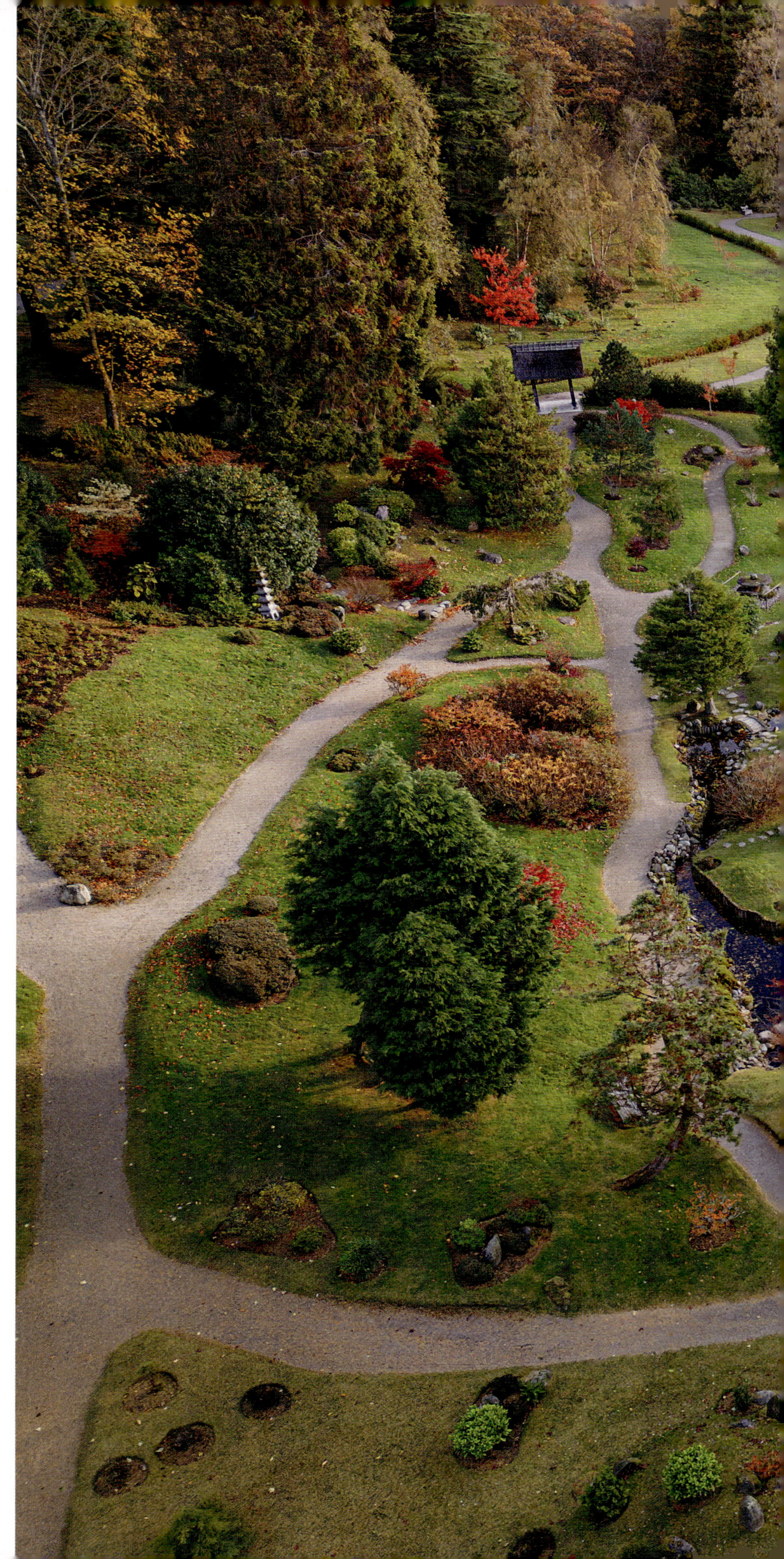

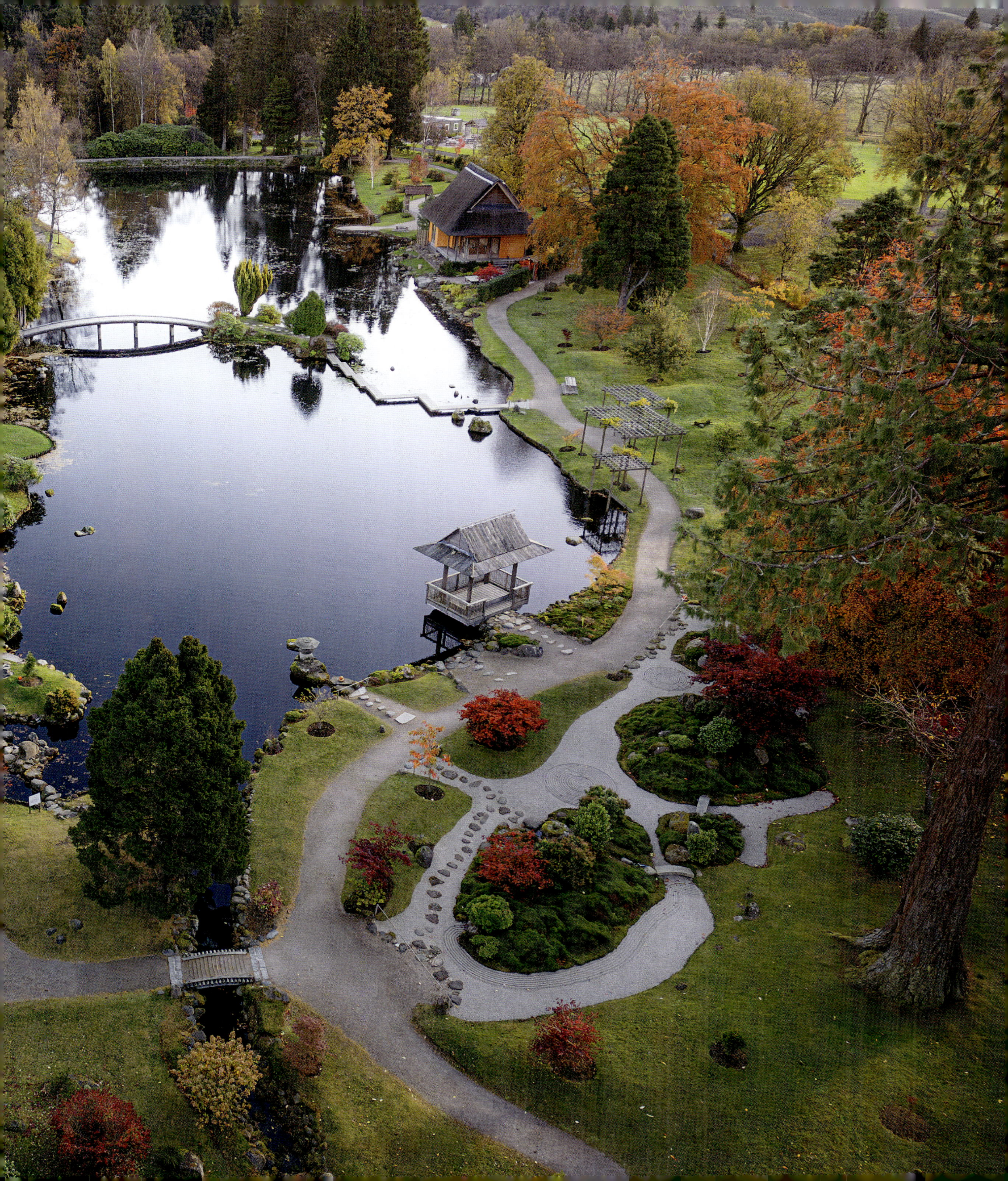

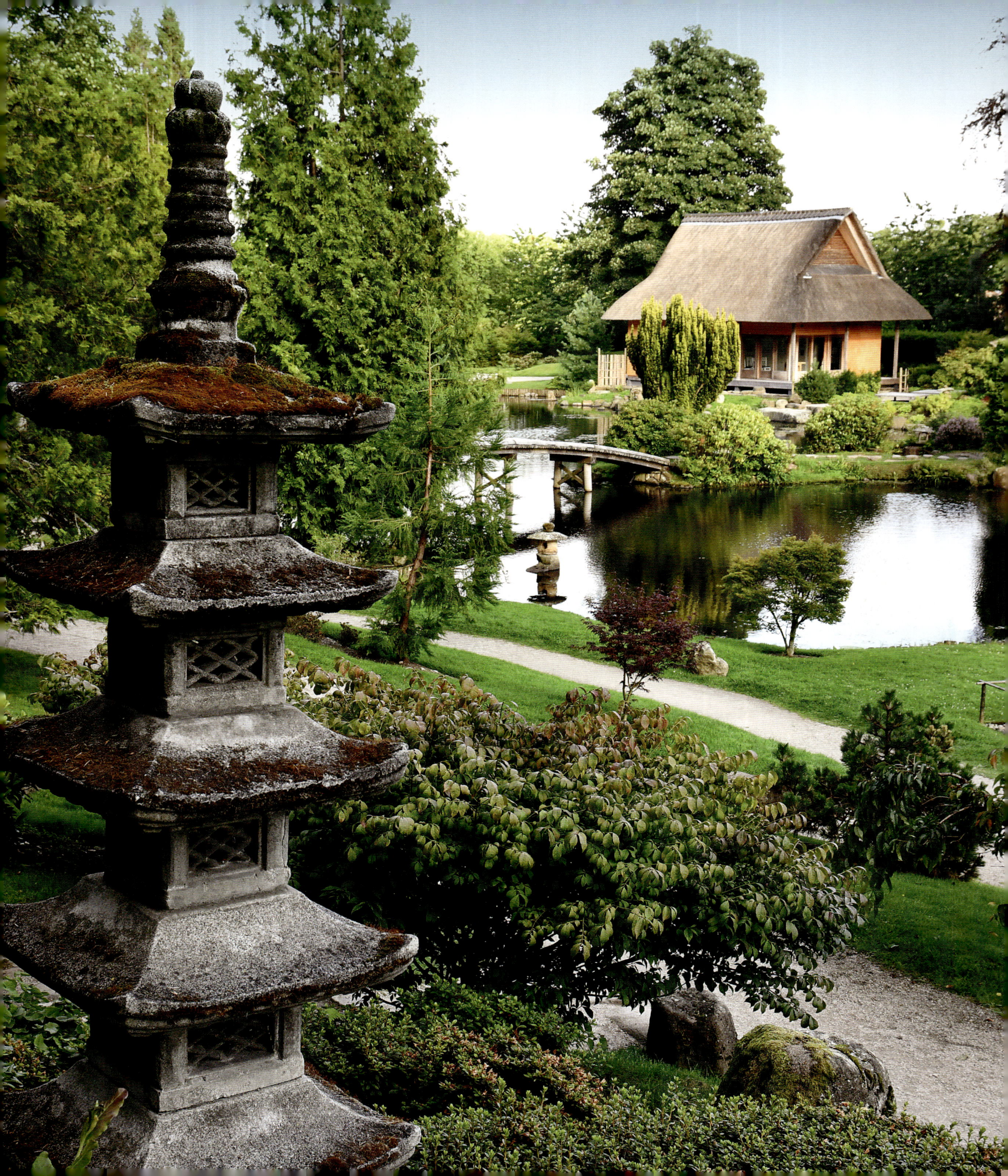

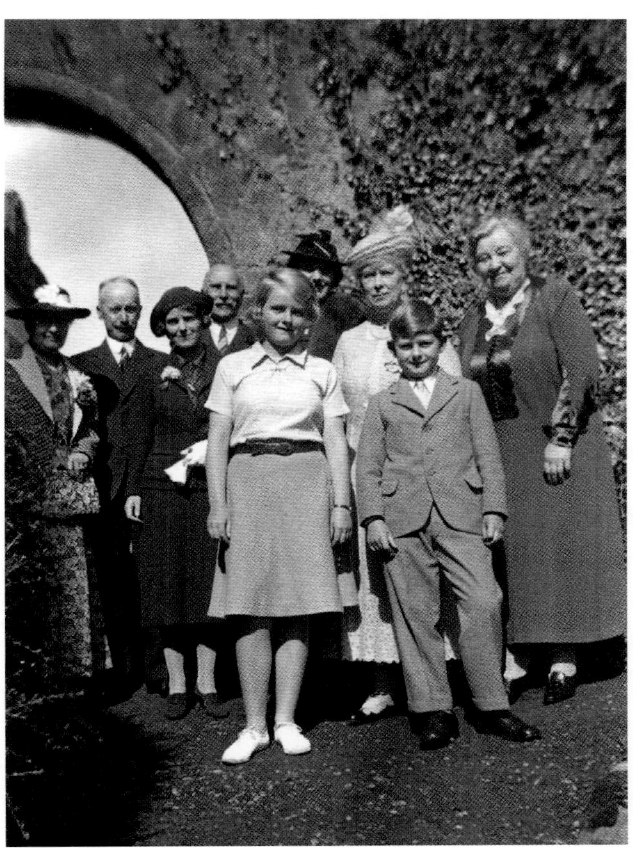

opposite South-facing view from the pagoda to the Garden Pavilion.

above Queen Mary visited the garden in 1937, while on a general tour of Scotland. She is seen in this group with Ella beside her (far right) and Robert Stewart and his sister Grizel standing at the front.

'When Isabella Christie died last January, setting forth on her final journey as calmly as on any of her adventurous wayfaring trips to the furthest East, there was more than a sense of loss among her friends. It was the end, the utter and complete end of an era; for in Miss Christie the past still lived along with the present. No one ever had a keener sense of that past or made history more living in interest; no one ever approached each day with a finer zest for life. In her strong and vivid personality, she was like memory incarnate, yet her memories never overshadowed present realities or new friendships. Only the old can link past and present, but only the very rare among the old do it well.'
— *From the obituary pubished in* Scottish Field *shortly after the death of Ella Christie in 1949.*

'Queen Mary, a friend of Grandaunt's, visited Cowden in 1937 and I remember showing her the collection of china, oriental *objets d'art*, and photographs of Ella's travels. We showed her the 13th-century tower and visited the long gallery, where I held the light cable for her to see properly. I remember Ella giving the Queen a piece of china. We drove to the garden on the recently constructed 'Coronation' tarmac road; the garden was immaculate.

   ... When my sister and I visited as teenagers, before and during the war, Grandaunt would explain all the meanings of the stones, the shrines and the lacquered lanterns. She explained how no one must meddle with the spring, which was the source for the water supply to the loch. She would sit in her revolving summer house, admire her creation and reminisce on her visit to Japan. It was never boring and made a lasting impression on me. For Ella, who had bravely travelled the world often in unchartered territory, this was her reward.'
— *Bobby Stewart in a recorded conversation with his daughter Sara, 2017*

Bibliography and Endnotes

Christie, Ella R., *Through Khiva to Golden Samarkand* (Seeley, Service & Co, 1925)

Christie, Ella R., *Fairy Tales from Finland. From the Swedish of Zacharius Topelius* (T. Fisher Unwin, 1896)

Christie, Ella and King Stewart, Alice, *A Long Look at Life by Two Victorians* (Seeley, Service & Co, 1940)

Duncan, Jane E., *A Summer Ride through Western Tibet* (Smith, Elder & Co, 1906)

Stewart, Averil, *Alicella* (John Murray, 1955)

NLS = National Library of Scotland
Letter-journal = Letters (some undated) written from Ella to her sister Alice in the form of a travel diary, which she would post whenever she could. These were later transcribed and bound into four leather volumes, now in the NLS.

1. Hand-written note from Ebenezer Henderson LL.D, 1875, still in the family's possession. Dr Henderson's major work was *The Annals of Dunfermline*.

2. *Ibid.* With edited punctuation.

3. Christie and King Stewart, *A Long Look at Life by Two Victorians*, p.32–33.

4. *Ibid.* p.37–8.

5. *Ibid.* p.20.

6. *Ibid.* p.58.

7. *Ibid.* p.45 and 46.

8. *Ibid.* p.51.

9. *Ibid.* p.54.

10. Stewart, *Alicella*, p.71.

11. *Ibid.* p.89.

12. *Ibid.* p.74, quoting from letter-journal written in the Howard's Hotel, Jerusalem on March 17, 1895.

13. *Ibid.* p.74, quoting from letter-journal written in the Grand Hotel Besraoul, Damascus on March 31, 1895.

14. *Ibid.* p.90.

15. *Ibid.* p.90.

16. An invoice in the NLS shows the charges for: 'Staircase & 55 panels including freight charges to Glasgow, 3 Carved Posts, 12 carved Corbels also packaging & post totalling Rupees 2160'.

17. Christie and King Stewart, *op. cit.* p.175.

18. *British Weekly*, date of publication not recorded.

19. Christie, *Through Khiva to Golden Samarkand*, p.5.

20. Stewart, *op. cit.* p.79, quoting from letter to Elizabeth Haldane written from Fonda Andrea de Bartolome Grau, Seo de Urgel, Spain on April 8, year not given but probably around 1896.

21. *Ibid.* p.80.

22. Christie and King Stewart, *op. cit.* p.67 and 68.

23. Stewart, *op. cit.* p.157, quoting from letter-journal written in Kandy, Ceylon on March 13, 1904.

24. NLS, letter no. 6/425, written to 'Bhamo' by M. D'Avera, agent to the manager of the Irawaddy Flotilla Company, in Rangoon, April 12, 1904.

25. Stewart, *op. cit.* p.162, quoting from letter-journal written in Kedah on May 18, 1904.

26. Letter-journal, June 6, 1904.

27. Stewart, *op. cit.* p.163.

28. *Ibid.* p.164, quoting from letter-journal written from the Grand Hotel, Simla on June 23, 1904.

29. *Ibid.* p.166, quoting from letter-journal written in Srinagar on July 26, 1904.

30. *Ibid.* p.166, quoting from letter-journal written on a boat on the Jhelum river on July 29, 1904.

31. *Ibid.* p.168, quoting from letter-journal written in Sonamarg on August 1, 1904.

32. *Ibid.* p.169, quoting from letter-journal written in Dras on August 3, 1904.

33. *Ibid.* p.168–69, quoting from letter-journal written in Sonamarg, undated but probably August 2, 1904.

34. *Ibid.* p.171, quoting from letter-journal written in Kargil on August 7, 1904.

35. *Ibid.* p.174, quoting from letter-journal written in Skarbichan Ladakh on August 27, 1904.

36. *Ibid.* p.175, quoting from letter-journal written in Paxfain on September 1, 1904.

37. Duncan, *A Summer Ride Through Western Tibet*, p.247.

38. *Ibid.* p.257.

39. Letter-journal, November 23, 1904.

40. Stewart, *op. cit.* p.184 and 185, quoting from letter-journal written on board SS *Heliotrope* on November 23, 1904.

41. Letter-journal, Islamabad, October 14, 1904.

42. Stewart, *op. cit.* p.186, quoting from letter written to Ella by Sir Frank Drummond in Goona, India on February 11, 1904 (now in NLS).

43. *Ibid.* p.188.

44. *Ibid.* p.190–92, quoting letter-journal written in a river-boat at Canton on February 20, 1907.

45. *Ibid.* p.193, quoting letter-journal written on board SS *Prinz Ludwig* on March 4, 1907.

46. *Ibid.* p.194, quoting letter-journal written from Hotel Astor, Shanghai, March 5, 1907.

47. Letter-journal, on board SS *Li Fong* on the Zangtzi Kaaig (Yangtse Kiang), March 12, 1907.

48. Stewart, *op. cit.* p.196, quoting letter-journal written in Peking on March 19, 1907.

49. NLS, letter to 'B' written by Ella from Astor House Hotel, Tientsin on March 26, 1907. 'B' is probably short for Baby, Ella's affectionate name for Alice.

50. Letter-journal, no date or place given, but by now they were in Mukden.

51. *Ibid.*

52. Letter-journal, no date or place given, but Ella was in Port Arthur at the end of March/beginning of April, 1907.

53. N L S, letter from Ella to Elizabeth Haldane, started at Port Arthur on April 3, 1907, and continued in Seoul.

54. *Ibid.*

55. N L S, letter to Elizabeth Haldane written by Ella from Miyajima on April 23, 1907.

56. Letter-journal, Imperial Hotel, Tokyo on May 4, 1907.

57. Letter-journal, Kindayu Hotel, Ikao, Japan, undated.

58. Letter-journal, Shamiogama, Japan, May 11, 1907.

59. Letter-journal, Tokyo, May 28, 1907.

60. Stewart, *op. cit.* p.207–08.

61. *Ibid.* p.208, quoting from letter-journal written on the Trans-Siberian Railway on June 15, 1907.

62. Christie, *Through Khiva to Golden Samarkand*, p.7.

63. The only hint we have of the identity of Ella's travelling companion is in a letter she wrote to Alice from Istanbul on March 7, 1910. In it, Ella refers to the Turkish passport people recording their names on arrival as Isabella Robert (her full name was Isabella Roberston Christie) and Marie Elysia, 'the last name never thought of,' as Ella writes. So who was her travelling companion Marie Elysia? An anonymous journal describing this trip is in the N L S. Somebody has noted at the front of it 'A M S diary,' but close inspection suggests that the hand in which the journal is written is not that of Ella's sister, Alice Margaret Stewart. Could it be that of Marie Adelaide Elizabeth Rayner Lowndes (Elizabeth being a variant of Elysia)? Mrs Lowndes, wife of the *Times* correspondent Frederick Lowndes and sister of Hilaire Belloc, was a good friend of Ella, a prolific novelist and a vice-president of the Women Writers' Suffrage League.

64. N L S, travel diary assumed to be written by 'M' (see footnote 63).

65. *Ibid.*

66. *Ibid.*

67. Stewart, *op. cit.* p.222, quoting from letter-journal written in Hotel de Londres, Tiflis on April 5, 1910.

68. *Ibid.* p.225, quoting from letter-journal written at Hotel de L'Europe, Baku on April 18, 1910.

69. *Ibid.* p.227, quoting from letter-journal written at the Grand Hotel, Askabad on April 23, 1910.

70. *Ibid.* p.229, quoting from letter-journal written in the Caravanserai, Bokhara on April 27, 1910.

71. *Ibid.* p.232, quoting from letter-journal written at the Grand Hotel, Samarkand on April 29, 1910.

72. Christie, *op. cit.* p.65–68.

73. Christie, *op. cit.* p.70–71.

74. Stewart, *op. cit.* p.244–45, quoting from letter-journal written in Khiva on May 3, 1912.

75. *Ibid.* p.245, quoting from letter-journal written in Khiva on May 4, 1912.

76. *Ibid.* p.247–48, quoting from letter-journal written on board s s *Carmania* on February 21, 1914.

77. *Ibid.* p.254, quoting from letter-journal written at the Savilla Hotel, Havanna on March 25, 1914.

78. *Ibid.* p.255, quoting from letter-journal written at the Greenewald Hotel, New Orleans on April 3, 1914.

79. *Ibid.* p.257.

80. *Ibid.* p.258, quoting from letter-journal written at the Utah Hotel, Salt Lake City on May 12, 1914.

81. *Ibid.* p.261, quoting from letter-journal written in Cold Spring on June 12, 1914.

82. Stewart, *op. cit.* p.262.

83. *Ibid.* p.272.

84. *Ibid.* p.277, quoting from letter to Elizabeth Haldane written at Cantine des Dames Anglaises, Bar-sur-Aube on June 4, 1916.

85. Letter-journal, Bar-sur-Aube, December 3, 1916.

86. Stewart, *op. cit.* p.280, quoting from letter-journal written in Bar-sur-Aube on October 16, 1916.

87. Stewart, *op. cit.* p.xvi–xvii.

88. Christie and King Stewart, *op. cit.* p.188.

89. Stewart, *op. cit.* p.304.

90. Christie and King Stewart, *op. cit.* p.234.

91. Stewart, *op. cit.* p.210, quoting from letter-journal written at the Imperial Hotel and Villa, Kyoto on May 3, 1907.

92. N L S, letter to Ella from Josiah Conder, written at 25 Mikawadai Machi, Azabu, Tokyo on May 9, 1907.

93. Tachibana, Setsu, 'The "Capture" of Exotic Natures: Cross-cultural Knowledge and Japanese Gardening in Early 20th Century Britain,' a paper published in the *Japanese Journal of Human Geography* 66–6 (2014).

94. Christie and King Stewart, *op. cit.* p.234.

95. Nakanome, Taki, *Omoide no ki* (Osaka: Kyoei Shinbunsha Shuppankyoku, 1954). (Taki Handa's memoir, written in Japanese.)

96. *Ibid.*

97. N L S, letter to Ella from T. Imai, written at the antiques shop Daidokuya, Kyoto on October 13, 1908.

98. N L S, letter to Ella from Suzuki, written at 8 New Oxford Street, London on October 6, 1925.

99. N L S, letter to Ella from Suzuki, written at 14B Morat Street, Brixton, London on December 18, 1933.

100. N L S, letter to Ella from Suzuki, written at 14B Morat Street, Brixton, London on December 28, 1932.

101. N L S, letter to Ella from Suzuki, written at 14B Morat Street, Brixton, London on January 24, 1933.

102. Stewart, *op. cit.* p.216.

103. Interview with Margaret Borland of Dollar, whose father was Ella's chauffeur.

104. Quote from a conversation between Bobby and Sara Stewart, recorded in 2017.

105. Stewart, *op. cit.* p.213.

Author's Acknowledgements

To start with, thanks go to my sister-in-law Sara who had sufficient faith in me to ask me to write this book and illustrate it with my photographs in the first place. Her energy and drive have been nothing short of extraordinary (shades of Ella) and I am hugely grateful to her as it's been a fun and interesting project.

Secondly, to my daughter Molly who has accompanied me on Ella related travels and with whom I spent many happy hours seeing who could get the best photograph of the garden in it's various lights and as it changed during the restoration process. Husband David and daughter Iona have also had to put up with a certain amount of agonising and being woken up horribly early in the morning when the light is good for photography, so I'm grateful to them too.

Next, to Mary Miers who knocked my rough prose into the polished end result that is between these covers and picked up numerous inconsistencies that had got into the narrative along the way on account of my enthusiasm.

Then to Robert Dalrymple for winnowing the photographs and arranging the book to look as enticing as it does. It was his inspired suggestion to start each chapter with a photograph of Ella's garden.

Finally, to Ella Christie herself for giving me huge amounts of raw material to work with. Women like her don't come around too often and I would imagine that if she could see her garden now, in all it's splendour and giving so much pleasure to so many visitors, even she might be amazed.

Lucy Stewart

A Word of Thanks

Ella Christie fascinated me as a child. An adventurous grandaunt who was as interested in the people she met as she was the scene in front of her. Her stories were legendary amongst friends and fellow travellers. She was clearly intelligent, mastering Swedish to translate a book on fairytales, but it was her lifelong curiosity and a desire for authenticity, which inspired me. A prime example being the Japanese Garden, created at a time when few people would have known if it was a mere pastiche or the real deal.

The garden I inherited bore little resemblance to what visitors now enjoy and I will be forever grateful to Professor Fukuhara who rose to the challenge to restore Ella's vision. Justin Rose's timber structures never cease to lift my spirits, the Garden Pavilion and Christie Lodge are beautiful additions by Hugh Shaw Stewart.

The generous grants awarded to the charity and the kindness of individual donors made the 10-year restoration project possible.

I am also extremely grateful to my father for entrusting the garden to me, to my mother for her unwavering support and to my sister-in-law, Lucy, for her dedication in writing this definitive biography of Ella. A huge thank you to the gardeners and volunteers who continue to toil in all weathers and to the indoor team who calmly manage the site for the benefit of all.

We all have one thing in common –
we love The Japanese Garden.

Sara Stewart
Chairwoman, The Japanese Garden at Cowden
SCIO Charity no.SC045060